Benjamin Lan
a Life in Physics at MIT

Benjamin Lax - Interviews on a Life in Physics at MIT

Understanding and Exploiting the Effects of Magnetic Fields on Matter

Donald T. Stevenson (Interviewer)
Marion B. Reine (Editor)
Roshan L. Aggarwal (Editor)

CRC Press
Taylor & Francis Group
Boca Raton London New York

CRC Press is an imprint of the
Taylor & Francis Group, an **informa** business

eResources: The link to download the eResources that accompany the book is: www.crcpress.com/9780367313500

CRC Press
Taylor & Francis Group
6000 Broken Sound Parkway NW, Suite 300
Boca Raton, FL 33487-2742

International Standard Book Number-13: 978-0-367-31350-0 (Paperback)
               978-0-367-31352-4 (Hardback)

Visit the Taylor & Francis Web site at
www.taylorandfrancis.com

and the CRC Press Web site at
www.crcpress.com

Printed in the United Kingdom
by Henry Ling Limited

We dedicate this book to our wives:
Marjory Stevenson, Kathleen Reine,
and Pushap Lata Aggarwal

Contents

Summary of the Life and Career of Prof. Benjamin Lax

PROF. BENJAMIN LAX (1915–2015) was born in Hungary and immigrated to this country in 1926. The Lax family settled in Brooklyn, New York. Ben attended public grammar and high schools in Brooklyn, graduating in 1936 from Boys High School with awards in mathematics, science, and languages. During the academic year 1936–1937, he attended Brooklyn College, where he received first prizes in two mathematics competitions. In 1937, he entered Cooper Union Institute of Technology in New York City, where he held the Schweinburg Scholarship for four years. In 1941, he graduated *cum laude* with a bachelor's degree in mechanical engineering and with honors in mathematics.

He then took a job with Curtiss-Wright Corporation in Buffalo, New York. He left that job after several months, and returned to Brooklyn to pursue graduate education in mathematics. In the fall of 1941, he took a job as a mechanical engineer with the US Army Corps of Engineers, inspecting tugboats in New York Harbor. Also that fall, he took several night courses in applied mathematics at Brooklyn Polytech and New York University. He married Blossom Cohen in New York City in February 1942.

In the summer of 1942, he began graduate work in mathematics at Brown University, but this was interrupted when he was drafted into the Army in August 1942. He underwent basic training at Camp Upton, New York, and at Fort Monmouth, New Jersey, after which he completed

Officer Candidate School. He was assigned as a radar officer to the Radar School at Harvard University and MIT, and later to the MIT Radiation Laboratory, where he led a team that rapidly developed and deployed an advanced radar set.

After the war, in 1945, he joined the US Army Air Corps Cambridge Field Station. He left the Army in February 1946 and consulted for Sylvania Electric Products in Boston for six months before beginning graduate school in physics at MIT in October 1946. Three years later, in 1949, he received his PhD in plasma physics. From 1949 to 1951 he carried on postdoctoral research in microwave gas discharges in the Geophysical Directorate of the Air Force Cambridge Research Laboratories.

He joined the Solid State Group at the newly formed MIT Lincoln Laboratory in November 1951. He advanced rapidly. He became Leader of the Ferrites Group in 1953, Leader of the Solid State Group in 1955, Associate Head of the Communications Division in 1957, Head of the Solid State Division when it was established in 1958, and Associate Director of Lincoln Laboratory in February 1964.

His groundbreaking research on resonance phenomena in solids in magnetic fields at Lincoln Laboratory led to his conceiving the idea of a new high-magnetic-field facility. In the late 1950s he assembled a team to prepare the necessary plans and proposals, and led the successful efforts to get the new national magnet facility funded. He served as the first Director of this new facility, the MIT Francis Bitter National Magnet Laboratory, from 1960 to 1981. Under his leadership, and with support from the US Air Force Office of Scientific Research, the National Science Foundation, and other agencies, this unique laboratory became a vibrant research facility that drew local, national, and international scientists to conduct broad research efforts at high magnetic fields in a wide variety of materials, including semiconductors, semimetals, insulators, ferrites, superconductors, gaseous plasmas, as well as biological matter.

He relinquished his Associate Directorship of Lincoln Laboratory in May 1965, and in that same year was appointed Professor in the Department of Physics at MIT, where he supervised 36 MIT doctoral students over a 30-year period, while also continuing as Director of the Magnet Laboratory until 1981.

He became Director Emeritus and Physicist of the MIT Francis Bitter National Magnet Laboratory in June 1981, and Professor Emeritus in the

MIT Department of Physics in 1986. He remained an active, enthusiastic, and valued consultant to MIT Lincoln Laboratory and to Raytheon Corporation until 2006.

He made significant and lasting contributions to solid state physics and engineering, as well as to the physics of both solid state and gaseous plasmas. He innovated new resonance phenomena, including cyclotron resonance, to determine the basic energy band structure of semiconductors. He pioneered the field of magneto-optics to further elucidate the fundamental energy band properties of semiconductors. He made important theoretical contributions that led to the demonstration of the first semiconductor diode laser in GaAs. His seminal basic and applied research in semiconductor physics and engineering, which included quantum electronics, provided a foundation for the development of semiconductor technology.

Among his awards was the prestigious Oliver E. Buckley Prize for Condensed Matter Physics from the American Physical Society in 1960 for "his fundamental contributions in microwave and infrared spectroscopy of semiconductors." He was elected to the American Academy of Arts and Sciences in 1962 and to the National Academy of Sciences in 1969. He was a Fellow of the American Physical Society (APS), a member of the APS Council, and a member of the Executive Committee of the APS Solid State Division. He was awarded an honorary Doctor of Science degree from Yeshiva University in June 1975. In 1981, he was a named a Fellow of the American Association for the Advancement of Science, and was awarded a Guggenheim Fellowship in Mathematics.

He was an associate editor for several journals, including *Physical Review*, *Journal of Applied Physics*, and *Microwave Journal*. He was the author or coauthor of nearly 300 journal articles, and coauthor of the classic book *Microwave Ferrites and Ferromagnetics* (McGraw-Hill, 1962).

"He who controls magnetism will control the universe."
Paraphrased from a Dick Tracey cartoon, 1932

"He who masters magnetism can bring enormous blessings to mankind."
Benjamin Lax, "Science and Magnetism," University of Utah Frontiers of Science Lecture, 16 December 1969

Timeline for the Life and Career of Prof. Benjamin Lax

Born   December 29, 1915, Miskolc, Hungary
Died   April 21, 2015, Newton, Massachusetts

Education

1941	BS, Mechanical Engineering, Cooper Union Institute of Technology
1941	1st Lieutenant, Officer Candidate School, Fort Monmouth, New Jersey
1949	PhD, Physics, Massachusetts Institute of Technology

Positions

1936–1937	Student, Brooklyn College
1937–1941	Student, Cooper Union Institute of Technology
1941	Curtiss-Wright Corporation, Buffalo, New York
1941–1942	US Army Corps of Engineers, New York City
1942	Summer graduate student in mathematics, Brown University
1942–1946	US Army Air Corps, Radar Officer, assigned to MIT Radiation Laboratory
1946	Consultant, Sylvania Electric Products, Boston
1946–1949	Graduate student, Department of Physics, MIT
1946–1951	US Air Force Cambridge Research Laboratories, Staff Member
1951–1953	Staff Member, MIT Lincoln Laboratory
1953–1955	Leader, Ferrites Group, MIT Lincoln Laboratory
1955–1957	Leader, Solid State Group, MIT Lincoln Laboratory
1957–1958	Associate Head, Communications Division, MIT Lincoln Laboratory
1958–1964	Head, Solid State Division, MIT Lincoln Laboratory
1960–1981	Founding Director, MIT Francis Bitter National Magnet Laboratory
1964–1965	Associate Director, MIT Lincoln Laboratory
1965–1986	Professor of Physics, MIT
1981–2015	Director Emeritus and Physicist, MIT Francis Bitter National Magnet Laboratory
1986–2015	Professor Emeritus of Physics, MIT

Selected Participations

1957–1959	Associate Editor, *Journal of Applied Physics*
1960–1963	Associate Editor, *Physical Review*
1959-1974	Associate Editor, *Microwave Journal*
1963–1967	Member, Council, American Physical Society, Member, Executive Committee, American Physical Society, Solid State Division
1964–1981	Member, IEEE-APS-OSA Joint Council on Quantum Electronics
1966–1968	Chair, IEEE-APS-OSA Joint Council on Quantum Electronics
1970	Chair, Organizing Committee, 10th International Conference on the Physics of Semiconductors, Cambridge, Massachusetts, August 17–21, 1970
1970–1981	Member, Solid State Science Panel, National Research Council

Selected Awards and Honors

1957	Fellow, American Physical Society
1960	American Physical Society, Oliver E. Buckley Condensed Matter Physics Prize
1962	American Academy of Arts and Sciences, Member
1964	Cooper Union Distinguished Professional Achievement Award
1965	Citation for Outstanding Achievement, US Air Force Systems Command
1969	Member, National Academy of Sciences
1969	Gano Dunn Medal, Cooper Union Alumni Association
1970	Outstanding Achievement Award, US Air Force Office of Aerospace Research
1975	Honorary Doctor of Science degree, Yeshiva University
1981	Guggenheim Fellowship in Mathematics
1981	Fellow, American Association for the Advancement of Science

Family

1940	Becomes a US citizen, June 8, New York City
1942	Marries Blossom Cohen, February 11, New York City
1948	Son Daniel R. Lax born
1950	Son Robert M. Lax born

Acronyms and Abbreviations

AC	Alternating Current
AFCRL	Air Force Cambridge Research Laboratories
AFOSR	Air Force Office of Scientific Research
APS	American Physical Society
CCNY	City College of New York
CERN	*Conseil Européen pour la Recherche Nucléaire* (European Organization for Nuclear Research)
DC	Direct Current
IEEE	Institute of Electrical and Electronics Engineers
IF	Intermediate Frequency
IRE	Institute of Radio Engineers
LED	Light Emitting Diode
LORAN	Long Range Navigation
MIT	Massachusetts Institute of Technology
NACA	National Advisory Committee for Aeronautics
NASA	National Aeronautics and Space Administration
NMR	Nuclear Magnetic Resonance
NRL	Naval Research Laboratory
NSF	National Science Foundation
NYU	New York University
OCS	Officer Candidate School
ONR	Office of Naval Research
OSA	Optical Society of America
PPI	Plan Position Indicator, a type of radar display
PX	Post Exchange
RF	Radio Frequency

RLE	Research Laboratory of Electronics, MIT
SDI	Strategic Defense Initiative
SQUID	Superconducting Quantum Interference Device
TEA	Transversely Excited Atmospheric, a type of laser
TR and T/R	Transmitter/Receiver
WKB	Wentzel–Kramers–Brillouin

Introduction

THIS BOOK PRESENTS A series of autobiographical interviews that Prof. Benjamin Lax (1915–2015) recorded in 1998–2000 on his life and career in physics.

These interviews took place at the MIT Lincoln Laboratory in Lexington, Massachusetts. The interviewer was Dr. Donald T. Stevenson, a career-long colleague and close friend of Ben's. The interviews were recorded onto 19 audio tape cassettes, with a total recording time of 20 hours. Don subsequently transcribed the audio tapes into Word files, resulting in a 300-page raw transcript. In 2005, Ben and Don donated the tapes and the raw transcript to the Niels Bohr Library & Archives at American Institute of Physics Center for the History of Physics in College Park, Maryland.

The interviews are approximately chronological, although there are occasional "flashbacks" to earlier events. The interviews begin with Ben's earliest memories of his childhood in a village in Hungary, and extend to his Emeritus years. The interviews contain both personal and professional reminiscences.

Chapters 1–10 in this book contain edited selections from the original transcript, representing about 150 pages of the original 300 pages, all in Ben's own words.

We have gently edited and annotated the selections in Chapters 1–10 of this book. Most of our edits were done to improve clarity and readability of what Ben was trying to express. We spell-checked and fact-checked Ben's text. We found that Ben's recollections of names and events were quite accurate. We added footnotes to give the reader background information on the many events, dates, meetings, publications, and people that Ben mentions in the interviews. We omitted duplicative text. Sometimes we melded parallel accounts of the same event so as to capture as much detail as possible. In some cases we rearranged portions of the text to accord better with the chronology. Occasionally, we modified words or sentences to more clearly convey Ben's stories.

In some places within the text, we added editorial or otherwise helpful comments. These are shown in italics and are usually contained in square brackets.

These autobiographical interviews are more like monologs rather than a traditional question-and-answer interview, so Chapters 1–10 contain only Ben's words. We omitted the occasional questions from the interviewer.

Numbers enclosed by brackets, such as [6.17], appear throughout the text and refer to items in the Notes section. For example, [6.17] refers to note 17 for Chapter 6. Alternatively, alphanumerics enclosed by brackets refer to items in other sections of the book. For example, [T34] would refer to a thesis listed in the section Theses Supervised or Mentored by Prof. Benjamin Lax. Similarly, references such as [P7.26] and [B11] would refer to items in the Selected Publications and Biographical References sections, respectively.

Reference to the interview tapes and raw transcript

"Benjamin Lax memoir [sound recording], 1998–2000," Niels Bohr Library & Archives, American Institute of Physics, One Physics Ellipse, College Park, Maryland 20740.

CHAPTER 1

Early Years in Miskolc, Hungary, 1915–1926

Taunted blind lady throws stone at young Ben Lax

My first recollection of my childhood goes back to when I was four years old. That must have been the summer of 1920, right after World War I ended.

There was a blind lady in front of the gate to our house, an iron gate. Our house had a garden which went up a cobblestone slope to the houses that were further back. She was in front of the house, and a bunch of hoodlums, who were children and were gentile kids, were throwing stones at her because she was half-blind. She kept cursing them and cursing them. She was a Jewish lady. I happened to be just coming home. Even at four my parents let me run around.

I was just coming home from a friend's house down the street when I passed her. She thought I was one of the hoodlums, and she threw a stone and hit me in the back of the head and knocked me out. Apparently some adults saw the incident and carried me to my house.

Nothing serious happened, but I remember this incident very clearly, and I remember these children taunting and tormenting this poor semi-blind lady. She could see shadows. She probably had cataracts.

Ben's home and environs in Miskolc

My father was Louis Lax and my mother was Amelia (Grosswirth) Lax. We lived in this house in Miskolc. The name of the street was *Szeles Utza*. At that time, I think my father may have been in business, which ultimately failed. But he became a wine salesman and made a decent living.

We lived in a house owned by a Jewish man. There were three families that lived there: his family, mine, and another family who owned a lumber yard and had the best apartment. We had an apartment with two rooms, a big kitchen, a large foyer, and, on the ground floor, an open covered area like a big porch. Often in the summer we used to take tables out there and eat there. We always had a maid, some peasant girl, who slept in the kitchen.

This place was very interesting because we lived in the middle house, the landlord lived in the next house, and just beyond there were the barns for horses and cattle which he owned. He owned milk cows. He also had another barn opposite this barn. There was a big yard that had chickens. Beyond the barn there was a big area, a big garden, where they grew vegetables. In front of our house there was an open area, part of which was the cobblestone walk. Beyond that there were trees, mulberry trees, which I used to love to climb. We were the only young children in these three houses. The lumber yard owner had two daughters, who were a little older than me. They were probably in their early teens. All three families were Jewish.

None of the houses, certainly not our house, had toilets. There was an outhouse where you went to do your thing. Regularly, gypsies used to come with a wagon that stank to high heavens, and we used to empty this sludge underneath. We would be there, smelling all miserable, but this is the way people lived at that time. This was in the early 1920s. They took the debris away. They probably used it as fertilizer. They had a big tank, and they filled this tank up and they took it away. It was very interesting.

Once you went down the cobblestone walk, on either side of the walk were gardens with trees … very interesting for us children … and then next to our house beyond the fence was a monastery. You could never go into the monastery. There were tall fences around there, and when you went out on the street there was a sidewalk in front. About a block away were my friends, and the father of one of my friends ran the grocery store. His name was Paul. The other friend's name was Mickey. We all were the same age, and later on, when we got older, we went to school together. We were in the same classes.

There were some nice houses along there. In one of them was a reasonably well-to-do businessman who had a beautiful daughter, whose name was Magda. I always admired her at a distance. I think she was about my age.

My recollection is that right in front of our house was a big open area called the *Buzater*, which essentially is an open area. Regularly, about once a month, the peasants used to come there during the summer with their wagons and horses to sell vegetables and all sorts of goods. This was an exciting time. We watched the horses. We always admired the horses. In the 1920s you never saw automobiles except on rare occasions. Most of the roads were dirt roads.

On our street, on the opposite side beyond the *Buzater*, opposite my friend's house whose father owned the grocery store, was a woodwork shop, a big shop, where they used to make furniture. I used to love to watch them. It always fascinated me. Beyond our house, beyond the garden, past the gardens of the monastery and the vegetable garden of our landlord, were open fields where we used to play soccer with the other children. So this was a very interesting area.

When you went away in the other direction on *Szeles Utza*, you went up a hill and then, when you turned right, you went up another hill. My grandfather's home was up there. If you kept climbing that hill, there were many mulberry trees that I used to love to climb as I got a little older.

We used to visit our grandfather, my maternal grandfather. They lived there with their daughters, my mother's sisters. He had a tavern. Peasants and others used to come and drink. He also had a lot of wine. He had a wine cellar, just beyond the house under the hill, just beyond the house. It was a very interesting rural area.

The next interesting incident that I remember was when I think I was about five years old, four and a half maybe … it was sometime after the blind lady's incident … airplanes flew overhead. In Hungarian we used to call them *repulo*. I remember standing there with my father. My father told me these were Russian planes. The reds were trying to infiltrate into Hungary. In fact, they came at one time and confiscated my father's cow, which he had for the children, which he kept in the stable. They took the cow that provided milk for us. So those are my earliest recollections.

Father saves Ben and brothers from a mad dog

I used to go to the synagogue on Saturday with my father. We young children would go and pray, and then around noon the services would be finished and we'd walk home. The synagogue we went to was not far from our school. We'd cross this little bridge that was over a little river between the main street and the school. Then we'd go across this big field, which was called the *Buzater*, where peasants used to bring their wares and sell their goods and products. In fact, that was also the area where during the summer the circus would put up its tents. It was a big open area in front of our house and in front of the Roman Catholic Church or the Greek Catholic Church, whatever it was.

So, this time of the year, I think it was in the fall, this big field was wide open. The circus and the peasants would come in the summer. I think this mad dog incident happened in the fall, approaching winter. As we were walking there, my father said, "There's a mad dog." The dog was heading directly toward us, and you could see the froth on his mouth. It wasn't a big dog, and somehow my father hurried us up and we got out of the way.

It was a scary thing because mad dogs attack people. There were occurrences in Europe at that time of rabies. So we were kind of scared, and fortunately we weren't too far from our home, and there was a metal gate there, although I think this dog could have gotten through it. We closed it.

Apparently the dog went by us. That was one of the scariest things I ever saw, because he was heading straight toward us, but somehow my father managed to get us away from it. We all were pretty mobile. I think it was myself, and my two other brothers. I don't think my oldest brother was with us, but Ernest and I think Dave were with us.

Begins kindergarten at age four and a half

At age four and a half, I was scheduled to go to kindergarten. I remember walking with my older sister, Ella, and crying all the way because I didn't want to go to school. I loved my freedom and the ability to roam around and visit my friends.

But I went to school, which was right up the street, just beyond the grocery store and my friend's house. I entered kindergarten without any incident. I don't recall anything exciting there. I suppose, after I got used to it, I enjoyed playing with the children because I was very friendly and sociable.

At that time I could only speak Hungarian. My father could speak Yiddish, but he never did. I spoke Hungarian to my father and mother. My father was well-schooled in Hebrew. He went to yeshiva until he was a young man, and that was his primary training.

Visits to grandfather's farm and uncle's farm

My maternal grandfather owned a farm in Czechoslovakia just north of the border. Originally it was Hungary, but after World War I it became Czechoslovakia. He also had a tavern where he entertained and sold liquor. He had ducks and geese and all the other things, so they made a living.

His brother also had a farm, and he was much better off. He was my godfather. He had a much bigger farm, called Jagfala, which was just south of the border. He had an orchard, he had wheat fields, and he had horses, three large horses. One was a gray mare that used to pull a two-wheeled carriage. The other two were work horses. He had two beautiful Shetland ponies that also were work horses that pulled the plow driven by his son, whose name was Dezso, which is the equivalent of David in American. My brother David's Hungarian name also was Dezso. He was a strapping young fellow who used to do all the farming.

They used to go to the orchard in the fall. I never saw that, but I visited the orchard. They picked the apricots and they made liquor. They had a still. They also made cheese because they had quite a few cows. They were truly good farmers, and he was well off. He also owned some real estate in Miskolc City, so he was considered to be wealthy. Every summer, from the time I think I was seven, I would spend the summer there, at their house, and I would have the freedom of the village … this was a peasant village … and I would just roam around the area and I would go visit my uncle who was plowing.

As I went through the village, I remember the peasant children would always taunt me as being a Jew and chase me and I ran like hell. My Aunt said: "This is a boy who never walks." So that's how I spent my summers.

Watches peasants partying on Sunday, their day off

On Sundays, the peasants came in from the surrounding farms. During the week the men tilled the soil. The women, the young women, used to work in different households as maids, to help with the children and everything else in the house. We had one of those, a very nice-looking young woman.

On Sundays, the men would be dressed up in their fancy outfits. These were the colorful traditional outfits of peasants, with fancy hats with a feather in them. And these young maids would get the Sunday off. They'd all go parading around in the main streets. It was very colorful. And, of course, they'd get together.

It turns out that our particular young woman had gotten together with one of the men and had a child by him. In fact, she had to work to support the child. They never got married.

Most of these peasants were uneducated. I don't think they even went to high school. They came from the farmlands. They often didn't own the farms. Some of the wealthier people owned the farms, and they just merely helped to till the soil.

My uncle, my godfather, who lived just below the Czechoslovak border, had such a farm. He used to hire these peasants to cut the wheat, and take in the fruit from his orchards. He used to make brandy. So all these peasant people would live in these villages.

This was a village north of Miskolc, and during the summer I would go there. At the end of the summer the peasants would reap the wheat, cut the wheat and stack it up, and put it into these machines that separated the straw from the wheat. But on Sunday these guys would really dress up fancy.

The land was usually owned by others. For example, my uncle owned a lot of land. By European standards he was a wealthy man. He even owned a townhouse in our town. He was a shrewd man. He also owned a bar. The peasants would come in on Sunday and drink, and gypsies would come in and play music, and it was a gay time. I remember I used to peek in to see these grown men whooping it up. But their children used to chase me all the time. Once they caught me and they started whipping me. So apparently these peasants were brought up this way, to be anti-Semitic.

Enters first grade at age five and a half

At the age of five and a half I started a school that was further away, about a half-mile or so away from my house, through the main thoroughfare or the side streets, which is the way we usually went. This school faced the stream that separated the business area from the school. This was a school run by the Jewish community, which was a secular school and where we started first grade. I went up there up to fifth grade.

Up until the fourth grade, I never bothered paying much attention, but I learned. In fact, they used to bring my father in and complain that Bela, that was my Hungarian name, was always daydreaming. But, nevertheless, I learned, and in my fifth year I became much more interested, and I guess I began to enjoy school. But up to then, I hated school. It just kept me from all the other things that I loved to do, roam around, play soccer.

Further down the street from this school was a Hebrew school, so after the regular school, around three o'clock, we went to the Hebrew school for a couple of hours to study the Bible in Hebrew, translated from Hungarian, and so on. This went on until 1924 when my father left for America.

Father leaves for America in 1924

He left for America in 1924. He had made up his mind that's where he wanted his family to live. He had wanted to go earlier but my mother objected. But he made up his mind that Hungary, Miskolc, was not a good place to bring up children. There were too many gangs roaming around, beating up children. My older brother, who was a tough kid, used to go around with brass knuckles to protect himself. I don't know what incidents he had, but apparently he came through OK.

When my father left, my mother, of course, was left with six children [1.1]. There were four boys and two girls. I was the fourth child. My sister Ella was older, and my brother Erno was older than her, about three and a half years older than I was. Then there was my oldest brother, Alex, the black sheep of the family, who was the tough guy. There were two younger children, my brother Dezso, David as we called him later, and my sister, Ilona, who changed her name to Eleanor.

Those were the six children. Most of us were happy children, and certainly the five of us got along very well. It was Alex who sort of ignored us because he was the eldest, and we ignored him. We kept clear of him. He had a temper.

Death of older brother Erno

Things went peacefully until about six months after my father left. My brother, Erno, was caught by a gang of kids. These were gentile kids who jumped on him, beat him, and kicked him in the kidneys. A few months later he started having trouble. After a while, he couldn't go to school and was bedridden. He was a marvelous, bright young kid. He had a tremendous sense of humor. He made all the children laugh. He even made the adults laugh.

In my opinion, he probably was the brightest of the lot. I never realized I was bright until later, but there's no question that I think he was even brighter than me. He was very talented. It was tragic. A year after he got kicked, they had to take him to Budapest to a hospital and he died there. The whole family was tragically affected. My father wrote a letter back and said that he thought the best of the lot died. I'm sure Erno was the brightest.

Studies Hebrew

During that time, as I went through Hebrew school, an elementary school, I was precocious. One of the Hebrew teachers decided to teach me at the age of nine how to read the Torah.

We always went to synagogue on Saturday with my grandfather, to a private home, part of which was converted to a synagogue, to pray. There they read the Torah during the prayers. At the end of the prayers there was a special passage which anybody could read, even a child. At the age of nine, I learned to read this, and this is done with intonation. The Torah, which is read in the temple, has certain symbols that tell you how to read this in a singsong. Apparently I had a good ear for music. I learned all this under the tutelage of this teacher, and to my grandfather's great surprise, I volunteered to read it at the week that this particular passage was due, and I did it. He was very proud.

My mother wrote my father, who was so pleased he sent me $3 for me to spend. I didn't spend it. I gave it to my mother. Three dollars was a lot of money. She could use it for the benefit of the family.

Trouble on the farm causes a hasty exit

When I was nine and a half, I went for my summer vacation to my godfather's farm as usual. One day there were these peasant children who were tending geese. One of them was a slightly older child. There were three of them. They were tending the geese and they cornered me. The older one started whipping me with his whip. At that age I used to have a temper. I got very angry. I had something in my hand, I think it was a piece of metal, and I just hit him over the temple. I think he fell down and I ran away. I didn't even see if I had done any damage, but apparently it stopped their ganging up on me. This was late in the afternoon or evening and, of course, I hid in the house.

I was frightened because I didn't know what I had done, but apparently I was told that I had injured him pretty seriously in the temple above the eye, and I think they were ready to lynch me. So that night, my family packed all my goods, and we took the gray mare with the carriage, and they put me on the train, and I guess they notified my father by telegraph or something. But early in the morning, I got on the train and I left for Miskolc. So that was the end of my summer vacations.

Trouble with boy next door

That brings me to another incident. There was a gentile boy next door to us who used to terrorize me for years, and every time he came, I ran. I don't know why I was afraid of him, but apparently he intimidated me.

But that summer we were playing soccer in the *Buzater*. He came along and he decided he was going to beat me up. I decided for once I'm going to stand my ground. Instead of his beating me up, I wrestled him to the ground and I started pummeling him. As was usual in wrestling in Hungary, you put your thumb in his mouth and make his mouth bleed. That was the end of that incident. He never bothered me.

But there was a fallout to that. Apparently he had an older cousin who had watched this fight, and he thought I ought to be whipped. This guy kept chasing me until we left for America later in August. This happened in July. This is the kind of atmosphere we lived in.

Mayoral election troubles

When I was about seven or eight, just before my father left for America in 1924, there was a big election in the city. A Fascist and a moderate gentile ran for mayor. The Jewish inhabitants were allowed to vote. And they swung the vote to the moderate. Then the adults who were associated with this Fascist, his followers, went around beating Jews, adults, in the city. I remember my father sent the maid to bring us home from school because they were afraid they would beat children too.

The maid came and picked us up at school. This was around midday. We walked home along the side streets. As we approached our house, on this big area which I called the *Buzater*, we saw in the middle of it a gang of hoodlums hitting one of these Jews with a beard, with a long coat and a black hat, one of the orthodox Jews. They were beating him. I guess they were very upset because their man lost the election. That was one of the incidents, among others, that I think influenced my father to leave Hungary.

Visits Tokaj with father

My father was a wine merchant, so he'd go to Tokaj [1.2], which was not far from our city. You go by train. Maybe it was an hour or so, or a couple of hours by train. Just before he left for America, he decided to take me and my brother to Tokaj, just for a trip. So we went with him on the train, and we got off. While he was doing his business, we were watching the peasants jumping in the big vats, crushing the grapes with their bare feet. This is a sight I saw many years later when I returned to Hungary. It was quite a sight.

Then one of the peasant boys came along and threatened to beat the two of us up. Of course, two of us were there and he had no chance. But this was typical of what happened in Hungary. I mean, these children, I don't know whether their parents influenced them that it was legitimate to beat up Jewish boys. Anyway, he couldn't do anything because we were two wiry, strong kids. We didn't, let's say, beat him up, but we held him in a head lock, and that was that.

But that happened time and again. Not just there, it also happened in America, as I'll tell you later on. I've had at least three incidents in my lifetime where some boys picked on me because I was a Jewish boy, particularly if I forgot to take my yarmulke off my head. You know what a yarmulke is - a little cap. But it wasn't as frequent as it was in Hungary.

Enjoys playing soccer

Miskolc was a very pretty city. In the middle of the city you had a park with a big hill, almost like a mountain, called *Avas*. It would have made a good ski slope. We used to go up there for picnics.

At the foot of this place there was a soccer field. Occasionally, we went there to watch the professional soccer players. There were two great soccer players on our city's team. One guy was a little stocky guy and he could run. The other fellow was a tall, handsome guy. He was what they call a striker today. So these were some of the things that children of my age (between six and ten) got excited about.

We played a lot of soccer in Hungary. We went to the fields behind the house with a soccer ball. Usually one of the boys would have one. We would just run around and kick it. There were, of course, older boys further down who would be playing soccer too. This was the favorite sport in Hungary, and most Hungarian kids who were athletic were very much interested in soccer. Soccer players were our heroes.

Tutored in 1924, learns German

When my father left for America in 1924, two years before we did, my mother decided to get us a tutor because we weren't doing that well in school. I wasn't very interested in school at that time. This tutor lived in a nearby housing complex. In Europe there were courtyards and there were apartments around the courtyards.

Instead of going to Hebrew school after our regular school, he would teach us both Hebrew and help us with our secular studies, like reading. He had a son who was studying English. He was a highly educated man. We were studying German. I could read German by that time, and that was one of the subjects I had to learn with him. And arithmetic, which I was good at anyway, so that was no problem. It was the other subjects I didn't care for.

He would tutor us in both Hebrew studies and Hungarian studies. It's too bad we didn't learn English at that time, which would have been an advantage. We could have started at a much more advanced level when we got to America. But I don't think my parents had that kind of foresight.

New interests: learning, agriculture, checkers

At that age, I don't think any of us were terribly interested in learning or studying, except for my last year, when I got interested in one of the subjects that was taught. The subject actually was agriculture. In Hungary this was a very important activity. I used to spend my summers on the farm, and apparently that hit the right button. So I was doing well in that subject, and I suppose by that time I began to realize it was important to learn.

The other thing I learned, about a year or so earlier, was checkers. I think I learned this from that tutor who was teaching us after my father left. Within a year I got to be so good that I could beat most of the adults that I played with. So that was another indication that not only was I competitive, but that I was fairly bright in mental activities as well as physical activities. Physically, I always was a spry kid who could run like a deer and who played soccer. Those were my primary interests.

Raises silkworms

The other thing that I also did during this period was raise silkworms. I think I did that even before my father left.

I would get the silkworms from the mulberry trees which were all over the area; in fact one was in our yard. I would climb up the tree, get the leaves, put these white silkworms on until they got to be very big, and then they'd form these little circular cocoons, and eventually the cocoon would burst and out would come a moth. A beautiful white moth, and they would lay black eggs. Then the black eggs would hatch and they would be little worms. At first they were black. Then, as they got bigger, they became these big white silkworms and they spun the silk. The cocoon that they spun was a silk cocoon. So it was a habit of mine and I loved to do that. I didn't unwind the silk. I just raised these things as a hobby.

Adept at mechanical drawing

Another thing I liked to do as a child, I think between the ages of seven to ten, before I left for America, was to draw. I loved to copy drawings of ships and other things. I was a fairly good artist, and I would do it scientifically. If it was a ship, I would measure out all the dimensions of the picture that I would copy, and then either I would duplicate it or increase the size. Multiply it. I used to love to draw, and I'll say more about that later on, because I think I continued doing these things.

Hebrew and singing

The other thing I also could do was sing. In our family, starting with my father, when we used to have meals, particularly on Friday nights, after our meal we used to sing certain traditional songs. This was on Friday and Saturday. All the children sang. In fact, I think every one of us had a good voice, maybe with the exception of Alex, our oldest brother. The girls and all the other boys could sing very well. Ernest was good at it and so was I. So that was another thing we learned.

Hebrew was my first learning. I can't remember when I couldn't read Hebrew. I think I learned to read Hebrew when I was three or four years old, because I remember reading the prayer at a very early age. So this is the kind of environment we lived in.

Active childhood in a rural environment

It was a very rural environment, with horses, cows. Sometimes when the cows were driven to pasture, we'd go down to the gates, which were closed, and this herd of cows was driven in front of our gate from the fields. Each group left its cow at its home. For example, they'd open the gates, and we'd of course hide, and the cows from our farm would come and by themselves would go up to the stables.

There was one other incident in my young life, when I think I was about seven or eight. The farm roof was about one story high, about 10 feet. Somebody challenged me that I couldn't jump off the roof, so I did. I didn't get hurt seriously, but I did feel a shock through my whole body. I guess I landed on my fanny. I was shook up, but otherwise I didn't get hurt.

This is the kind of child I was. I was a daredevil. I used to climb trees. I was a little roustabout climbing up. I was an active little boy.

Leaves for America in August 1926

Then came the time when we had to leave. That was August of 1926, two years after my father had left. We all packed up, the five children that were left, and my mother.

We took a train to Budapest. I guess about 100 miles. Miskolc is east, largely northeast of Budapest. And we arrived there, and my mother's sister lived there with her husband and two children who often visited us during the summer in Miskolc. We stayed with them for a few days. That was the first time I had ever seen Budapest.

For me it was a very impressive city. I mean, compared with Miskolc, which was a small town, although today it's a good-sized city. I had a cousin, I've forgotten his name, who was my age but about twice my height. He and I went swimming to a bath. You know, in Hungary you could go to a public bath. They had little pools. My cousin could swim, I couldn't, but these pools weren't that deep.

We went into the bath and he showed me around the city. I had a good time there. After a few days we left. We left, I think, late in the afternoon and we went through Vienna at night so I never really got to see what Vienna was like. Then we went through Switzerland and I think, I don't know how long it took, but we arrived the following afternoon in Paris.

In Paris, we were met by my mother's other sister, whose husband had left Hungary to live in Paris. He owned a business, a plastics business. He was relatively well off. He had a very nice apartment along the Seine. We stayed there for a few days, and we got to see Paris. We went to the gardens and so on. I don't think these were any of the famous places, but we spent a pleasant time there. Then it was time to go to Cherbourg.

We arrived at Cherbourg, on the coast of France. We went by train of course, and we arrived there, and I guess we stayed overnight there. Then we were taken by a barge out to a big ship which was the Cunard Olympic. There we transferred from the barge or tug, whatever it was, because the ship couldn't dock at the harbor. It stayed in the harbor, but the harbor wasn't deep enough. There were no wharves that could accommodate it.

We got on there, and we went third class, at the end, steerage. The whole family was packed in a little room with bunks. We were really very crowded. We began our journey to America.

Of course, we could never go to the middle-class or first-class areas. We were in the rear of the boat. In the rear of the boat there was a swing. I think that was the first time I ever got on a swing, and I was there all the time playing on the swing.

I think it took us a little over a week. The weather was great almost all the time. Apparently, being on the swing didn't bother me.

There was an old gentleman who had a chess set, and I started playing with him. I spent my time playing chess and amusing myself on the swing.

I didn't get seasick until the last day when it got rough, and I got very seasick. We were near the kitchen. I could smell the fish and I was very sick. I couldn't eat fish for years after that. The smell of fish – I still don't like it. It was terrible.

But I had a very good time, and I managed to amuse this elderly man. I don't know how old he was, maybe he was in his 50s. He was amused that I could play chess that well. I would beat him most of the time; occasionally he'd beat me, but we were about on equal footing. He couldn't believe that a ten-year-old boy could play chess that well. In those days, I had the interest and patience. Chess was new to me and it was a challenge.

CHAPTER 2

School Days in Brooklyn, 1926–1936

Arrives in New York Harbor and Ellis Island

After that stormy day, the following day, I guess it was late afternoon, we saw the Statue of Liberty. We came into New York and we got into one of the wharves. We were taken to Ellis Island.

Coming into New York Harbor and seeing the Statue of Liberty was a real impressive sight for us. It was twilight, I think, when we got to Ellis Island.

Takes subway to Queens with father and uncle

Soon my father and uncle arrived and we were put through the paces. It was nighttime and we were processed. Then for the first time in my life we got into a subway, which, when you went into Queens, went under the East River. My uncle lived in Queens. I was terribly impressed with the subway. The whole scene of New York was overwhelming.

We arrived in Astoria and we were taken home. Our home turned out to be an apartment inside a small synagogue in which my father was a sexton. He was the sexton of this small synagogue and the large conservative center next door to it. I guess the two were affiliated. He served both, but our living quarters were in the small synagogue.

Soon thereafter, my uncle came with his Buick and took us to his house that evening after we arrived, and we had dinner at his home. He had two children. He had two girls who were about the age of myself and my younger brother, David. We were about ten and nine years old. We had a nice dinner and we were very warmly received. Then we were taken home.

They lived maybe a mile away from us in Astoria. This was the Jewish Center on Crescent Street, which was a nice broad street. Right across the way from the center was a big apartment house, and there was a beautiful walk in front of the center, a cement walk, fairly wide.

Receives roller skates and a tricycle

The next day somebody gave me – I don't know where it came from – a pair of skates. I immediately proceeded to put them on, and in just half an hour I learned to skate on my own. Pretty soon I was buzzing around like I had been on them all those years. As I said, I was athletically inclined.

We lived there for about a year. Then shortly after we got the skates, we got a tricycle. You know, we never had anything like this in Hungary. So we got a tricycle, and my brother David and I kept going for hours around the block. We had such a good time.

America as compared to Europe seemed like, well, a Garden of Eden, you know, much more exciting and much more fun. We were there for about a month, August. Then in September school started. We should have learned English during that time, but we didn't. My folks spoke Hungarian, and we never went anywhere, really, except to our cousins who also spoke Hungarian. So we didn't have a chance to learn.

Starts school in grade 2B

Then they took us to school. The first day in school I was put into grade 2B. That was the second semester of the second year. The teacher was a big, intimidating lady who apparently wasn't impressed by my inability to speak English, and she said things to me in a tone of voice that scared the hell out of me. There was a little Italian kid who started making fists with his eyes, with the intent to beat me up or something. He scared me and I burst into tears. But I got over it.

So we would get math homework and we would get English homework. I was a whiz at math. She would say to me (later on I understood her): "How come you know math and you're so stupid you don't know English?" I mean, this was a grown woman, and I should have been given special reading, but apparently this teacher was incompetent. Anyway, I got through it that half-year.

There were several interesting incidents. This little bully, who was a little bigger than I was, after we got out of school one day, confronted me and wanted to beat me up. Well, I turned a trick on him. I beat him. I really whacked him. I mean, I just stood my ground. I wasn't going to take it from him. I thought that was the end of it.

We used to come home from school, and there was a lot where there was tall grass and there was a walkway in the grass to give you a shortcut. We used to take that shortcut. One day a pile of kids jumped on me, including this one. Apparently it was a little Italian gang. They must have been children of the Mafia! So they learned to do that. They started beating me, and I screamed like hell, and apparently an adult came along and they ran. But that's the way they did it. That was a scary moment, but that sort of thing happened, and so that was my first half-year in Astoria.

There also was a Hebrew school, so I went to the regular school, and in the afternoon I went to Hebrew school. My father, in addition to being a sexton, also taught Hebrew there. That was a good job. I don't know what happened, but he lost the job the following summer. Apparently he and the Rabbi didn't get along for some reason.

Finishes 3A and 3B in June 1927

At any rate, the second year I was put into a class with a very wonderful teacher, in contrast to the other one. Her name I think was Astair, and she taught both 3A and 3B, and my sister was in 3B. By that time, I could speak English a little.

So I was taking instructions for both classes. Sure, she would alternate between the two, but we were in the same room. By the time spring came (and I began to understand English much better), I had learned everything in both classes and that was it. I covered two classes at once. In one morning I could assimilate this learning. I wish I had the opportunity to do that later on. So in June of 1927, I was finished and I was ready to go into 4A.

Moves to Kelly Street in the Bronx

But then my father lost his job, and we went to live in the Bronx on Kelly Street. Kelly Street is located down in South Bronx. It's the same street that General Colin Powell lived on in that area. But at that time it was primarily a Jewish area, with synagogues and things around, and my father decided to send us to a parochial school on the other side of Bronx.

Even at that tender age of 11 and a half, my brother and I would take the subway to go to school. I attended grade 4A. First, in the morning, we would learn the Hebrew subjects, and then in the afternoon we would learn the secular subjects. You only studied maybe three hours instead of the regular five hours that you would spend in a regular school, and it was done in the afternoon.

So you would go to the school from nine o'clock until about six or seven. You got home at six o'clock or seven o'clock via subway. It wasn't a subway, it was elevated. So we learned to go there by ourselves, but today you wouldn't think, especially a little runt like me, to be let loose in New York on the subways all by myself. That was the situation then.

Interest in singing

About that time, my father realized that I had a good voice, and he took me to a synagogue, a big one where there was a choir, a beautiful choir, which sang wonderfully on Saturdays and holidays. There was a soloist who sang like an angel. (Apparently when I got to be a soloist, I did just as well.) I got very interested. I got very impressed by this young singer, and I wanted to do the same if I ever got a chance, which I ultimately did.

Father out of work in 1927

But my father couldn't get a job for six months. This was just before the depression, in 1927. That summer and all of 1927, he didn't have a job, and I don't know how we managed to make out. I guess he did odd jobs.

We lived in this apartment. We had a tenant because we needed the extra cash. He was an elderly Jewish man, who apparently was a widower. My mother was a wonderful cook, so he occupied one of the rooms in the big apartment and also was served a meal. So my mother was very busy. My mother was very hard working, a wonderful woman. She had a rough life.

My father also struggled. But at the end of the winter, around December or January of the following year, 1928, he got a job in Brooklyn as a sexton in one of the synagogues in Brooklyn on Sumner Avenue.

Attends yeshiva in Manhattan

My father enrolled my brother and myself in a yeshiva in downtown Manhattan. It was right off the Essex Street subway station. We walked about maybe a quarter of a mile or so, and it was near the Manhattan Bridge. In fact, you could see the Manhattan Bridge. I've forgotten the name of the street. It was not far from the building that housed the Jewish newspaper called *The Forward* that my father used to read all the time. It was a typical Jewish community down there.

We left early in the morning, and started school at nine o'clock 'til about 12, maybe with a half-hour break in between. Then we had lunch. Usually my father would give us some money to buy lunch at one of the local restaurants. They had Jewish restaurants. You could get a good cup of soup or a good sandwich for about 15 cents, and that was a meal.

We'd get maybe a quarter, with ten cents for carfare. It was a nickel to go on the subway from where we lived in Brooklyn down to downtown Manhattan, to the Essex Street station. We went over the Brooklyn Bridge, I think. There were three bridges across the East River: the Manhattan Bridge, the Brooklyn Bridge, and the Queensboro Bridge that led into Long Island. We went over the Brooklyn Bridge, and then we walked maybe a quarter or a half-mile to school, which was right at the Manhattan Bridge. That's when some of my interesting adventures began in Brooklyn.

The first adventure was when I got there, got to Brooklyn, and went out into the street, I was met by a bunch of boys. I think they were Jewish boys. One of them was a tough kid, and he said to me, "I can beat anybody on this block. You're going to have to fight me." So we fought. He, I think, beat me but not by much. Anyway, from there on we became good friends. But he had to establish that he was the king of the block, and it didn't bother me except that I wouldn't accept it at first. He was a little taller than I was and a little huskier than I was, and we wailed away but nobody really got hurt. But he realized that I wasn't going to knuckle down under him. So we had mutual respect for each other, and that was that.

Wins competitions in spelling and arithmetic

This was the first time I began to show excellence in school. In the afternoon, between four and seven, we had the secular studies. I was in 4B, which was the end of the fourth year, second semester of fourth year.

This teacher always had competitions in spelling and arithmetic. I constantly won. I always beat all the other boys. Now, how come in spelling? I mean, after all, I was at that time maybe about 12 and a half, just two years into America, but I did a lot of reading on my own. The first year I read in Hungarian. In fact, the first year I was here in America, I would read, for my age, around 30 advanced books like *Tom Sawyer*, *Last of the Mohicans*, *Uncle Tom's Cabin* in Hungarian [2.1, 2.2, 2.3].

Then, a year later, when I switched to English, I started thinking in English, counting in English. But in reading I was fairly behind. I started reading fairy tales and simpler things. I would buy books like *Tom Swift*, *Tarzan*, and read those things [2.4, 2.5]. So I would acquire a good vocabulary, and I have a retentive memory, so I got to spell pretty well, and I would beat all these kids. Arithmetic just came second nature to me.

Running races in Van Cortlandt Park

There was a playground there. Right after we had our lunch, we would go around, run, play soccer or whatever, right in the playground not far from the yeshiva. There was a supervisor there so nobody got into trouble.

At the end of the year, they decided to have an outing where all the kids would participate in races. Kids about my age or a little older. I remember I decided I was going to win the running race. So we started the race. This was in Van Cortlandt Park, up in the Bronx. There was a kid, a burly kid, about my size but a little huskier. When we got started, he elbowed me and I couldn't quite catch up to him. He just beat me by maybe a yard. He cheated, but anyway he was declared the winner. But I could run like a deer. And this was an enjoyable time.

Skips grade 5A

That teacher who had these competitions in class was very impressed by me, and decided to skip me. So I skipped 5A and he promoted me to 5B, and I wanted to go to summer school. He said, "No, if you go to summer school I won't skip you," which would have been a good thing for me to do.

So that summer I had off, and that's where my adventures in Brooklyn really began, because I had time to myself.

Builds scooter, enjoys Erector Set

Right near the elevated station that we used to take to downtown Manhattan, there was a big beautiful bank with a most wonderful sidewalk in an L-shape, and part of it was downhill. I used to go skating there.

Then I decided to take my skates apart, and I built myself a scooter with two boxes and, I guess, a two-by-four or something, and I would use it as a scooter. I would take it up to the top of this walk and come down with it, and I used to get around with this scooter. I would take my skates apart and nail them underneath. It made a wonderful scooter.

We didn't have money for anything, but I would make my own toys. I would find an old carriage and take it apart. I remember I used to pound on these things until I got the nails loose and then I would get some wood. You always could find some wood around. And I would make myself a wagon.

So I already began to show my ability to build things. In fact, I used to buy chocolate and we'd save the covers until we could get enough of them, both Dave and I, and I got an Erector Set. I used to assemble this. Make all sorts of things, not only the ones it showed in the book but ones that I decided to make on my own. I remember I had a new motor, and I guess I didn't understand electricity. It had batteries, and I think I put too much voltage on it. Eventually, I burned out the motor. But it was lots of fun.

I used to keep myself busy, and occasionally I would get into some sort of fight. When you went around, there were Jewish boys and there were some Italian kids. I think many of these Italian kids were Mafia members; they always fought dirty. One of them was about my size, very wiry, maybe a little older. He got me into a strangle hold and got the best of me. So, occasionally you would get into a fight, but for the most part I would try to avoid them. But that's the kind of place Brooklyn was.

Another school, more fights

After the summer of 1928, my father enrolled me in another yeshiva, which was closer to our home, but again you had to take the elevated, which was not far from the Williamsburg Bridge. There was the same routine. We started in the fall of 1928.

My first experience in the fall, the first thing, another kid in one of the other classes, a little bigger than I was, challenged me to a fight. And we fought like hell over the stairs and nobody won. I think one of the teachers finally stopped it. Of course, once again we became friends. I just wouldn't take any guff. I was always combative, not always to my best advantage, but most of the time I would undertake to fight somebody whom I thought I could at least keep up with. That was one incident.

Another incident: both my younger brother David and I went to this yeshiva. One day, I guess it must have been lunch-time or something, we were outside the yeshiva. Around the corner came a gentile boy, and he said, "I'm going to beat up you Jewish boys." He was a skinny guy, a little older, taller than I was, and he wouldn't have stood a chance against my brother and me. And I said to my brother, "I'm going to take care of the guy myself." Which I did. I got him into a headlock and gave him a few punches and that was that. But you probably didn't get the best of David and me. David was almost as big as I and very strong. At that time not quite as strong as I was, but we both were very good scrappers. That's the sort of thing we ran into from time to time.

Yeshiva studies, Hebrew, boredom

This yeshiva was an experience for me. We would sit there, again it was three hours in the morning with a break in between. During the break we would play soccer with a tennis ball down the street. Then there was lunch at 12 o'clock. We would eat our lunch and play soccer again. Then at three o'clock there was another break, and we were through with the Jewish study.

But we had essentially, with the break, maybe five hours. The best hours of the day were Jewish studies, and it was boring as hell. The Rabbi or the teacher would read the passage of the week. You know, every week you read a different part of the Torah, or the Bible. And it takes a year for you to go through this. That's what you read on Saturday in the synagogue. And we had to translate it. We had to translate it from Hebrew to Jewish.

When I first got started I didn't understand either. But, just with my retentive memory, I could repeat it. So I would be the first one to volunteer to repeat it, and then would daydream the rest of the day as each student would repeat it. This took the whole part of the morning. So maybe there were 20 students, so you repeated the same passage 20 times. What kind of pedagogy was this? You can see why I got bored with it.

Then, in the afternoon, we would learn some Hebrew grammar and other things. I did have a good teacher, a man named Greenberg, and he was very fond of me. Fortunately, I had him for several years. I didn't want to be promoted. I said that's enough. I mean, his teaching advanced with every year. In other words, in the first year I don't think we had grammar. But then we learned Hebrew and grammar and things at the end, and that I found interesting. You know, once I got to understand and learn the Hebrew words and the Hebrew language and the grammar, to me it was interesting.

Deposits money for teacher

This teacher liked me very much. This man apparently had a lot of real estate in addition to being a teacher. He would collect money, and he had to deposit it maybe once a week. He would give me the rest of the morning off and told me to go to the bank and deposit his money. He could trust me. I would go and deposit the money. At least I wasn't bored. He'd give me carfare. I think it was downtown in Manhattan.

So I would take the elevated on the Williamsburg Bridge, end up downtown, go there, and take my time before I got back. I got back in time for lunch in the afternoon. But that would be my morning about once a week.

Starts singing at age 11 and a half

A year before, just when I was 11 and a half, my father took me to a choirmaster named Machtenberg. My father wanted me to sing in a choir. And the choir leader decided to test my voice. He played "O Promise Me" on the piano. I looked at the music. Although I couldn't read music, I could read the letters. After one playing, I sang the whole song.

I had at that time such a retentive memory. In fact, I never learned to read music, because if I would hear a song once, then I could sing it. So he took me into the choir. This was the choir that previously had Jan Pierce as the soloist, who ultimately became a famous opera singer. He didn't have a soloist, so I became the soloist.

Sings at synagogue on Riverside Drive

He had a big choir. That fall his choir sang in a big synagogue or temple off Riverside Drive in Manhattan. A fancy place. There was a famous cantor, Vigoda, who had a beautiful voice. I was the soloist. My brother and I both became members of the choir. You know, Jewish people can't travel during the holidays, so I had to be in Manhattan. So my father arranged with one of the Jewish members of the temple, and we stayed with him. He lived not far from Harlem, in a stone home, a very nice one. That's where we stayed during the holidays while we sang in the choir.

We would walk around Riverside Drive. It was beautiful area. It was not yet occupied by the blacks. In fact, it was a nice Jewish community. That's where I think I. I. Rabi [2.6] and his parents lived, in that general area. This was a wonderful temple and was quite an experience. That began my singing career.

Joins new choir, prepares for bar mitzvah

After that experience, when I was 12 and a half, I was to be made ready for my bar mitzvah. My father decided to enroll me in a different choir, and to get a cantor who would teach me to sing the prayers that a cantor sings for the Saturday on my bar mitzvah.

Within six months I became a boy cantor. The cantor taught me all the things. He was trying to teach me to read music, but I never bothered to practice because I would pick these things up fairly easily.

In December, when my bar mitzvah took place, here on Sumner Avenue, I read the Torah. I gave a speech in English, a speech in Yiddish, and read the Torah. I'm telling you, I had a good voice. The ladies in the balcony were crying. I'm not joking. I was, well, you know how I am, I was a show-off and a showman. And I really outdid myself.

The next day we were supposed to have a party for me. And what happened? I came down with the flu. So everybody enjoyed themselves on Sunday, and I got from my uncles watches and things. But I was in bed. They had a party downstairs. There was a room in the synagogue for the party. So all the friends of my father and my uncles came. I had uncles from my mother's side, two of them, and from my father's side, his brother. But I missed my party.

After that, my father would schedule me to sing at various temples, and I hated that. I had no intention of becoming a cantor. It was fine for bar mitzvah, once, I thought, as a special occasion. I did that for several years.

In the meantime, every year I would sing in a choir. This cantor, his name was Smulikowski, a red-headed man, had a nice family. He lived maybe a mile away, and I would walk there to take the lessons. Then, the following fall, we would start practicing for the choir. He would rent a place downtown, and the choir would practice and I would practice. I became the soloist again for his choir. I sang with him for several years.

Weekly baths at the public bathhouse

We didn't have a bathtub in the Sumner Avenue apartment. My father would take us to a bath place about a quarter of a mile away every Friday afternoon, when he had time off from the yeshiva. So on Friday afternoon we went down there, my father and David and I. We would go to what's called a *mikvah*. This is a public bathhouse for Jewish people. I guess in the old days nobody had baths, so this must have been a tradition.

They had a hot pool, a sauna, and a cold pool. We would take a shower and a bath. We would do that every week. I guess on other days the women would do this, so my mother and sisters would go then. We would take a bath once a week. This is the way we lived. We were very cheap.

Father says *kaddish* prayers

We didn't have to pay rent, but my father made extra money on odd jobs, and he would get donations for different services. You know, it's a tradition that, if somebody dies, you would say a daily prayer, a *kaddish* as they call it, for a year. Most of these people were fairly religious, but they couldn't do that. They had to go to work. So they would hire my father to say the *kaddish* for them. They would give him a few extra dollars, and he would earn money that way. Then there would be weddings and so on.

Sings for weddings

Talking about weddings, I told you I had joined this choir with Machtenberg, I think when I was 12 and a half, before I started studying for my bar mitzvah with this other cantor. He would have his choir sing at weddings. In June of 1928, there were 30 weddings all over the city. He had taught me "O Promise Me," and I would sing "O Promise Me" at each of these weddings. Now he would get extra money for that, I don't know how much, but for each "O Promise Me" he would give me one dollar.

So I made $30 that June. I gave it to my mother. It fed my family for over a month. Can you imagine a 12-and-a-half-year-old boy earning such money? So I earned money for myself by singing over the years. I never was without a job. Unfortunately, I couldn't take a regular job like a delivery boy because we were religious and they had to deliver groceries on Saturdays.

Moves to Bedford-Stuyvesant at age 13 and a half

We lived on Sumner Avenue until I was 13 and a half. Then we moved to another synagogue on Gates Avenue, which was maybe a couple of miles away, in the Bedford-Stuyvesant. Right in the heart of it. At that time it was mostly Jewish and some Italians.

There again we lived in an apartment, this time not downstairs like the other two, but upstairs, on top of the synagogue. It was an apartment. That was a little better than what we had before. My father would get, I think, $10 a week for his sexton job, so he had to make extra money for the family.

Roller skating, playing on the roof

I was about 13 and a half when we went there, to Gates Avenue. I would continue going to the yeshiva. To save money, I would save a nickel, especially on good days, I would use my roller skates. The yeshiva was maybe three miles from my house. I would roller skate there and roller skate back in the evening in the good weather. I would save ten cents for myself, the amount my father gave me for carfare, and I would buy baseball gloves and other things. Both my brother and I would do it. I got to be pretty good at roller skating. I always had roller skates.

I would use the money to buy roller skates. In fact, at first I bought roller skates with four wheels. Then later on, I bought roller skates with two wheels. I was a very good skater. Imagine skating in those streets. Occasionally, I would hit a stone or something and I would fall and rip my pants or fall on my knee or elbow. One time I skinned my elbow, and I didn't treat it, and it got infected. I had to go to a hospital. They made me bathe it in boric acid and it got better. But that's the kind of a boy I was, a little skinny boy who would undertake all kinds of silly things.

Talking about adventures, on Sumner Avenue there was a tall synagogue, and behind the synagogue there was a fire escape. From there you could go to the roof of the next house. I would take a clothesline, lasso it up to the roof of the synagogue, which was maybe 20 feet, and I would climb up that rope. I guess I was light enough. I would be on the roof of the synagogue. Once my father saw me and gave me hell. But this was the kind of adventurous kid I was. I would use clotheslines and other things to amuse myself. We didn't have money for toys, for bicycles.

Sings in summer resort in Monticello

When I got to be 14 and a half, my choir leader and cantor (the cantor was also the choir leader) got a partnership in a summer resort in Monticello, New York. The deal was, I would sing for him for the holidays, and he would pay me half in cash, and for the other half, both my brother and myself would spend two weeks at his summer place. We did that for two summers, I think, when I was 14 and a half and 15 and a half. That was the first time I ever had a vacation in a summer spot.

One of these people who sang in the choir, I guess his father was a partner there, would drive us up. He used to be a truck driver. This young man also sang in the choir as a tenor. He would drive us up and we'd spend the two weeks there. They had a swimming hole there, and I finally taught myself to swim the breaststroke. I don't think I really knew how to swim, not until later on.

For the first time, I think, I really ate wholesome meals. I found out you ate cereal like oatmeal for breakfast. You'd have bread and butter and lots of milk. I was very skinny. In fact, one of the councilors there saw me. He said I must have had rickets because my ribs stuck out and I was very thin. I put on some weight. But I was small and skinny for my age. Much smaller than kids who were two or three years younger.

We played handball there. They had a handball court and I started to learn how to play handball. We had a good time. Those were two summers. During the summer, I would have these various projects. I would make things for myself.

Then I started the habit of spending the summer getting the latest popular books and reading them. Of course I used to buy *Tarzan* and all these other books [2.5]. Later, maybe not until I was in high school, did I start reading more serious contemporary literature. I liked to read. I would get these boy's books, *Tom Swift*, *Tarzan*, maybe some others, and I would read during the summer [2.4, 2.5]. That was one of my hobbies. I don't know how I spent the rest of my summer. There was a library not far from there. I would go to the library and I would get books, you know, that were, let's say, up to the level.

School routines

In the yeshiva, the secular school, they didn't really give me homework, didn't give me assignments to read, so I was retarded for my age and level of study because the yeshiva was not like the regular school. They just didn't have the time and, hell, you were in school from nine until seven and you traveled half an hour to get there, half an hour to go back. So from about 8:30 am to 7:30 pm you were away from home.

Then you came home, ate dinner, and sometime around ten o'clock you went to bed. You didn't have much time to do these things except the weekend. And on the holidays, there was nothing to do because the Jewish religion forbade you to do these things. This went on until I got to be 16. That was my last term in the yeshiva. At 16, I was in grade 8B, so I was going to finish regular school.

That fall, we needed money for the school for graduation. I would go around to all the delicatessens in Brooklyn and downtown and solicit money, because we would advertise in this yearbook. I got to be pretty good at it. I recruited most of the money. The owners of the delicatessens were very impressed with this little guy who would go around and con-vince them that they had to advertise. All the delicatessens advertised. I didn't tell one delicatessen that there was another guy. I was a little con man. But anyway, I got it. So that was an interesting experience.

Bored with school, wants to be a scientist, not a Rabbi or cantor

There was an incident during that time. I was put into a more advanced class. In fact, this was the second advanced class. After Mr. Greenberg, I was put into another class in the yeshiva. A man named Shapiro was the teacher. He was much more strict than Greenberg.

I would daydream. Or I would have a book underneath that I would read, and he'd catch me. These were English books because I was bored. He would rap my fingernails with a ruler. It really hurt. This would happen time and time again, but I would do it anyway. Or I would just simply sit there and daydream because I was bored.

Then I got into this last class with this tall guy, about a six-footer. I decided I had had it. I just stopped paying attention. I didn't volunteer, I didn't read. He used to call me stupid. One day, just to show him up, I listened to his reading and repeated everything verbatim. He was livid. He said, "You've been fooling me, you're not stupid." So he started swinging at me. He was going to hurt me. I ducked, he hit the wall and I ran out. I refused to go into class. And he complained to the principal.

So my father came down. He was asked to come down because I would refuse to attend that class. It was near graduation. We sat down with the principal. I told him I was 16 years old. I said this is not the kind of education I want. I don't intend to be a cantor. I don't intend to be a Rabbi. I'm going to be a scientist. I want to go to a regular high school. I'm not continuing here.

So they solicited a promise out of me. They tried to convince me to do otherwise because they realized that I was a bright boy, but I just had no taste for this. And so I promised my father. My brother Dave, who was younger than me, broke off a half-year before I did. He went to summer school and graduated at the same time I did, but he refused to go that half-year. But I was a more obedient boy.

But I told my father, "OK, I'll study with you on Saturdays." Can you imagine, here I am going to school and I have to study, so I'll try to keep up with my Hebrew studies, which I didn't. But I tried to avoid him. So when Saturday came and he came looking for me, I went to the Brooklyn Children's Museum or I was out of the house.

Begins Boys High School, Brooklyn

So in 1932, at the age of 16, I started Boys High School. They gave me a test I didn't know about, but later I found out that I was deficient in my English studies, in other words, the size of my vocabulary and the writing thing. I had never written a composition. But that was the end of my career as a Jewish scholar. I started high school.

At the high school, we went to an annex, not the main school for Boys High School, which was near us. It was the annex, about a mile away. We walked there. In the annex, we studied regular school. You went from eight o'clock until about three o'clock, with breaks in the morning and lunch recess. I took four main courses: civics, German, English, and math. English was my weakest subject. I had difficulty in trying to write compositions because I had never done it. I didn't even know what a composition was.

Wins German competition

But I did all right there. In all the subjects, except English, I was the top student, particularly in German. The German teacher liked me. She was very beautiful young woman and she had two projects. One was that you're going to have to give a declamation in German. You're going to recite a poem in a competition for the entire school. Kids who were interested in German and German studies participated. She gave me a poem [2.7] called "*Erlkonig*." It's a very sad poem. I memorized it. When the time came to recite it, I cried like a baby. It was so sad that I won the competition. Everybody was pleased and surprised, and my teacher was very proud because I was her star student.

Physical education, working out

First term, there was another incident. In addition to having these (four) subjects, we also had minors. One of them was physical education. We didn't have a gym there, but we'd get exercises, and there was a gym teacher. He had a chinning board on the door. The doors were very tall. In fact, they had to lift me up to the chinning bar. He decided he was going to see how many kids could chin and how much they could chin.

Well, I forgot to tell you that I used to go during the summer, before I started going to Coney Island, I used to go to Tompkins Park, which was about a mile away, and I would work out on the high bar and the other things. I had to shin up to the high bar. I would watch the older boys doing all these gymnastic tricks like the giant swing. And eventually I learned to do these things, but I started there. They had the monkey ladder.

I saw what nice muscles they had, and I decided I was going to have muscles too. Of course, I never got muscles, but I got to be pretty strong. So by the time I was 16, I was fairly strong.

Anyway, he decided to see how some of the boys could chin. Of course, I chinned more times than anybody. I chinned 12 times. And maybe none of them could even chin 10 times. In fact there was one boy, named Wilson, who was quite an athlete, who played baseball. He was much bigger than I was and more muscular, but he couldn't chin as much as I could.

So, again the physical education teacher was surprised. He said, "Well maybe that's because you're light." And I said no, that's because I exercise. I told him what I do. This is the sort of surprise I always had for some of my teachers. In other words, I could do things they didn't think I was up to. They always were surprised.

Summer at Ravenhall, Coney Island; much exercise

I was 16 and a half, and as I said, I would make a little money on the side. My choirmaster didn't have a position that summer and he gave up the resort. I decided, instead of going to Tompkins Park, I got a season ticket in Ravenhall, Coney Island.

They had these public pools with lockers. It cost you $5 to get a locker. You could go all summer, at any time of the day. It would take me a nickel by subway or elevated to go from my place. You walk about a mile and half to one of the lines. You make a change to get there, Coney Island. Then you walk another mile to get to Ravenhall. The walks were nothing.

So I got to this Ravenhall. They had high bars, they had gymnastic things, they had handball courts. So my summer routine would be, I would go in the morning, take a sandwich. First we'd play handball, then we'd go swimming. I wasn't the only one, some of my friends joined me from around town. My boy friends I'll tell you more about, some other stories.

And we'd go swimming. Then we'd eat lunch. We'd play cards. After cards, and after settling down, I would go on the high bar and we'd work out for a few hours. Then we'd play handball again. Then I finally would end up in swimming. So we'd work out maybe two hours in the morning, about three hours in the afternoon. It was quite a workout. We'd catch lots of sun. Sometimes we got burned, but after a while we got tanned. That was most of my summer.

We practically were almost an hour on the subway train. I couldn't just sit doing nothing, so I would take reading material. That's when I started reading *Anthony Adverse* [2.8] and some of the other books on my own. There were some very good books, contemporary literature that came out of the library. I was a compulsive reader.

Joins a new choir in the summer

I would do all that, and various other things, and that would be my day. And then, before I came home, there was one more thing. My choir leader was no longer in business, so I decided to enlist in another choir, in Sheepshead Bay, which was on the way home. You know where Sheepshead Bay is. Coney Island, then Brighton Beach, then Sheepshead Bay. That is where the Russians are now.

There was a nice Jewish center there, a new center, fairly well-to-do. And I trained with that cantor and his choir. Of course, I was already experienced and didn't have to train. I knew all the songs. So they were delighted. I was the soloist. At 16 and a half, I still had a good alto. I still hadn't reached puberty. Can you imagine, 16 and a half. That's how I sang in the choir. That was my summer. And then we started school.

Second year at Boys High School at age 16 and a half, with five majors

This time, I was in the regular school in Boys High School, in the main buildings. That was my first experience with running from one class to another instead of just staying in one classroom all day. And getting gym and various other activities, and in the homeroom, study periods. I think we started studying biology.

It turned out I was also a good biology student. Again, it was a question of memory. The biology teacher used to lecture to us slowly, and ask us to take notes and make drawings. And I would take notes. Then he would give us tests. The tests were exactly what was in the notes. And I had an almost photographic memory.

Gets 100s in biology

I got 100 in every biology test. I guess I got 100 for the term, an A. Of course, there was no question, I was the best student. The next term we had physiology. Again, it was a question of memory, and I found physiology very fascinating. Of course I was very good, also the best student in that class.

We had two football players, two Irishmen, boys. Boys! They looked like men. They had beards. Here I was, this little runt, just beginning to reach puberty, and here were these 6-footers who played football, one was the quarterback, the other was fullback, who liked me. They thought I was a genius, which I wasn't. They were very impressed by how I memorized everything. I would read the book and I would remember everything in there.

We studied algebra. Then the next year, when I was studying physiology, we started geometry, and my home room teacher was my geometry teacher, a wonderful teacher. I did very well in math.

German was one of my favorites, I did well in that. And I had a history teacher, Mr. Ester; oh, what a wonderful teacher. So I was getting very good marks. Nothing was below 90. English I struggled with, and the first time I got a B. My first quarter. I went up to the teacher. I said to her, "Look, I know I'm not as good," but I said, "I want to be an A student, I want to compete." I thought it was important, since I thought I was going to go to college.

So I told her she could give me extra help and so on. I ended up with an A. But they give you a numerical mark. Whereas I would get 99 or 98 in most of my other subjects, I would get maybe 91, 92 at best, in English. So that was my weak subject. I got better at it. It took time though. I enjoyed high school. I thought it was very exciting. I enjoyed trying to get good marks. I was very competitive.

Tyrannical French teacher prompts switch to Spanish

The first year you could only take four majors. But if you did very well, they allowed you to take five majors in the second year. So for my second year, I had to add a second language. I chose French. We had a lady teacher who was a tyrant. She'd criticize everything. "You're not pronouncing right," she would yell at you. I said to hell with this, I'm not going to take it.

So I went to the principal. I said I don't want to study with this teacher. She had a reputation. I said I'll take another language. They changed me to Spanish. I was two weeks behind but I caught up soon. They gave me a book and I started studying. "*Pablo es un nino Espanol*" was the first sentence I read in the book. I became very good at it. Later, after the second year, we also had Spanish.

I was a cocky little kid. Some of the teachers didn't like me. This Spanish teacher was a lady teacher. One of my best friends, Stan Manson [2.9], who was also a very good student, was a year behind me, and he also took Spanish. We were in the same class. He was a good-looking, tall boy, and I guess the teachers cottoned to that too.

At the end of that semester, there was no question I was the best Spanish student in the class, including Manson. There was a medal up for grabs.

I guess that was the end of the two years. I only took two years of Spanish. You could take the final if you wanted to, since I got very high marks in the midterm (I think I got 100). Or, you don't have to take it, you get your A anyway. But if you took the final, each teacher could pick a student to represent the class to vie for medal. She picked Manson.

I went to her and I said that's not fair. I think I'm the better student. I said, you know what, you don't have to give me permission, I can take the exam even though I don't have to. I took the final and I got 100. He got 98. So her prize student didn't win, I did.

But there was another guy who also got 100, and he was a Spanish major. They could have given us two medals. They decided to give it to him because he was a Spanish major. But anyway, I won the thing, I made my point. I don't remember what she gave me, but I'm sure she gave him a higher mark. I probably ended up with 98 or something like that.

But, you see, this happens all the time. Even in school, teachers pick favorites on appearance, personality. They didn't like cocky kids. And I suppose I was a little bit of a showboat and I spoke up, I volunteered too much. They would say, "Give the other students a chance," or something like that. Well, I was enthusiastic. I wasn't bad, I was just enthusiastic, and I wanted to show up a little. But it never was done in bad taste.

Trains improperly for distance running, injures leg muscle

When I was in high school, I decided I was going to train myself. This was in the fall of 1933. I was starting my sophomore year in the fall.

I decided to train myself, not knowing really the proper way to do it. I would get up in the morning. I would go out, and I decided I was going to be a distance runner. I would run a half-mile or a mile. I never warmed up, and I kept doing it every morning for about two months, probably starting in September and going into October. I could run a half-mile every morning, even longer at almost full speed. I guess I got to have pretty good stamina.

What I didn't realize was that I was doing it wrong. You're not supposed to run without stretching and warming up. One day, I think at the beginning of November, on a Sunday, we decided to play touch football, which we used to do all the time on the streets of Brooklyn. The quarterback threw the ball to me, which I caught. I started sprinting. One of my friends, who was a pretty fast runner too, I thought was catching up to me. So I speeded up. All of a sudden I couldn't lift my left leg. Just couldn't. I quit, went home. My hip ached, and my father decided to take me to a hospital.

They took an X-ray and found out that the muscle had torn off, had chipped the bone that connected the tendon to the hip bone. So they hospitalized me. For six weeks I was lying on my back with my legs stretched so the muscle would relax and this thing would go into place. Sure enough, it went in and it healed by itself. The next six weeks they put a cast on me. The cast was all the way from the hip down to the ankle. I lost a term of high school because of this. That was very disappointing. I loved school, and already I was behind.

That's when I began to read avidly, and I got everything that I could. They could have sent me my homework, but nobody did. So I lost a term. But I did a lot of reading. I did so much reading that, all of a sudden, I needed glasses. In fact, the next term I found out in class that I couldn't see the front board. So I had to get glasses. That was a traumatic experience.

By the time it was January, I was already walking around with my cast. I could walk anywhere. At the end of January, just before the second term began, they took off the cast. My knee had been locked in all that time, and when I tried to bend it, there was the most excruciating pain.

But I recovered quickly. In about a week I was walking. By the time spring came, I was in school and doing fine walking, and my leg had recovered. That was the term by the end of semester I was getting 100 in my physical class. Well, it didn't take me many months to recuperate. Apparently I was young and strong.

And then there was a very interesting thing. I had a gym teacher, and in spite of the fact that I had this injury, I was pretty agile. He was very impressed. This was a very nice teacher. We also took hygiene. In other words, in addition to the gym class, they had a hygiene class, which meant you learned about nutrition, the body, the male organs, the female organs. You learned about nutrition and what was a balanced diet -something I didn't know much about.

Anyway, the gym teacher was very impressed by what I did. After a while, when he'd mark my paper and I would get 100 on it, he would let me mark the rest of the papers. And then I did very well in the chin. By the end of the semester, I was running again like a deer and I would run around the track. He gave me 100 in gym. I couldn't believe it.

Now this was one of the mature teachers who thought, gee, here's a kid who tries very hard. He and I got along. He understood me. He didn't resent my cockiness, although I don't think I was all that obnoxious. I got better as I got a little older.

But, anyway, I got 100 in gym. I mean, that's unheard of. I wasn't a perfect athlete, but he figured, you know, this kid had just recovered and look at what he's done in a period of about four months. After that I was perfectly fine. I could do everything.

Excels in math

This is the period, when I went to Boys High School, which I think were some of the happiest years of my life. The summer I would spend at Coney Island, and I would spend the evenings practicing for the choir.

Then I reached my junior year. In the first semester of my junior year, I took intermediate algebra. I was really fascinated with it, and it turned out I was the best student in the class. Actually, the class was composed of all the best students. It was an honors class. Still, even though it was an honors class, I think I was the best math student in the class.

At the end of the term, the math coach, who coached the math team, called me into his office and told me I should go out for the math team the following fall. He gave me some books to study. Now, after intermediate algebra you usually took trigonometry. But he suggested I take advanced algebra instead. That would help to prepare me for the math team since I had never sat and trained with them.

So in the fall, when school started, every lunch time I would spend in his class where all the students who were good math students were supposed to train for the math team. They would give us problems and we had to solve them. I got to be pretty good at it, but not yet experienced enough to make the math team. In any event, I enjoyed it and I joined the math club, too.

The following semester, which was the end of my junior year, I still wasn't yet ready for the math team, but I became a member of the math club, and they had a geometry math competition which I won.

It was the end of my junior year that there was a citywide competition, which included algebra and geometry, and he put me on the team to represent the school. I did pretty well, but maybe came out fiftieth out of several hundred. But that was the year that Richard Feynman [2.10] was a senior at Far Rockaway High School in Far Rockaway, and he came out first, way ahead of everybody. Apparently he had a tremendous score on that competition. And of course he was a year ahead of me. He was a senior, so he was better prepared for it.

The following year I came out number two. In fact, I thought I would come out number one. It turned out that a bright kid in my school, a year behind me, beat me to number one. But I guess we were close. In any event, that was quite a competition.

In my last year, I became a member of the math team and its star player. I used to score the most, and my team would beat all the other schools in the city, including Stuyvesant High School, which was our strongest competitor.

Photographic memory

They used to give us notes in civics the day before. And I would just read the notes. One day she sprang a test on us. And I happened to have just read the notes once, and apparently I remembered everything. I was the only one who passed the exam in the class. Normally they warn you, you're going to have a test, and then you study the night before, but I happened to have read the notes and with just one reading I was able to pass it.

Another time that I found out I had a good memory was in elementary school. There was a guy who had a photographic memory. One day he read the page of a book and I said, "I think I can do that." So I got the book and I concentrated as hard as I could, and indeed I did remember it, I could duplicate it. I don't know whether I could do it as well as he did, but I was able to duplicate that.

In fact, that guy became a good friend of mine. The way he got to be a good friend, we got into a quarrel one day. He was a slightly huskier boy, but I bloodied his nose. After that we became good friends.

He was also a good student. In fact, the boys in this yeshiva, most of them were bright, and the bright boys used to sort of form a clique. We played soccer and we studied together, and we just became pals. But that's when I realized I had a good memory, when I duplicated this feat of his. I discovered that's the way you do it. It's a question of concentration.

That's why, when I would take an exam, I could see the page in front of me, the equations, particularly in the sciences and the mathematics. I suppose you're born with this. But you cultivate it, too. You have to practice it. So this young fellow who demonstrated this capability was responsible for getting me to learn how to do that.

It was just a question of concentrating and, I suppose, being born with a good memory. But concentration is a very important part of this. In fact, when I used to study in high school and in college at home, I would concentrate so hard, you could burn the house down. If I were to read a novel or something, I was oblivious of what went on around me. It was a question of getting yourself into it.

Solves math problems in his head

I learned to solve math problems in my head because on Saturdays – at that time I was still an observant Jew – you couldn't even write. So when I was in the temple and I got bored listening to the prayers or to the reading of the Torah, I would daydream. But instead of the usual daydreaming, I would start solving math problems in my head. I could see all these equations and the solutions. It was great trait, and I became very adept.

Winning math team in senior year

This trained me very well for that math team. So when we competed, I'd have the problem done before they tapped on the table and you could start writing. They'd give you five minutes or three minutes, depending on the complexity of the problem.

And I would usually score a perfect score, up until the last competition in my senior year. The star of the Stuyvesant High School team was a fellow named Paul Marcus [2.11], who I later got to know here in Cambridge when I went to graduate school. Paul Marcus and I were tied for all of the sessions except the last one. I had perfect scores, but in the last session, I think I missed two of the problems. He got a perfect score, so I came out number two. But our team won, nevertheless. We were a better team than Stuyvesant. So it was a successful year. I made number two on two of the competitions citywide, so this was quite an achievement, and it certainly taught me a great deal of mathematics.

Self-taught calculus

My coach used to give me extra books, books in English. In winning the competitions, instead of taking the medal, I would ask him to buy me a book instead. So he bought me some of these English-language texts that were in advanced geometry, in algebra, advanced algebra, and I taught this to myself.

At the end of my junior year, that summer, as I was going to Coney Island, I bought myself a calculus book by Granville [2.12]. I taught myself calculus, both differential and integral calculus, just by reading the book going down to the pool, Coney Island, and back. I had completed the entire book by the end of the summer. Did most of the problems in the book. In the back there were a whole bunch of integrals. I had learned to go through the procedure to solve them. I think, at that time, I could remember every integral in that book. I remember helping some of my friends who were a year ahead of me in City College and who were studying calculus. I would tutor them and help them on how to solve some of the problems. I got to be that good by my senior year.

Sings in Carnegie Hall with Molly Picon

I sang in Carnegie Hall at the same time Molly Picon (1898–1992) did. They had some sort of a benefit. My father arranged it. I would be one of the entertainers. I don't know what songs I sang. Molly Picon is a very famous Jewish singer and actress. She played in "Fiddler on the Roof." She was the yenta. She had a good voice. So I sang on the same stage with her.

Sings in synagogue, earns money as holiday choir singer

When I was a boy cantor, my father made arrangements for me to sing in different synagogues. I would sing the services. I hated it but I did it. I'm not sure I got paid for it. I didn't get anything, but he did it because he wanted me to be famous or whatever. I don't know. You know how parents are when they have a talented child.

He was more ambitious for me as a singer than I was. Once I got into high school I stopped doing that. I sang in a choir for the high holidays because I needed the extra money. For that I got paid. I would get about $40 for singing on Rosh Hashanah. I would use that to buy myself clothes and other things. I matured before my younger brother, Dave, and I was still singing at 17 in a choir. I think I just about reached puberty then. Dave was 15. He had already changed his voice. He couldn't sing in the choir as an alto. So that year he didn't sing. When we both sang, we both would get paid. But that year he didn't make money, so I used my money to buy him clothes and myself clothes. I remember we'd buy a pair of pants and we'd buy a leather jacket. A leather jacket for both of us lasted for years.

Takes exams for Regents and Pulitzer scholarships

In the spring I took both the Regents exams, but I wasn't eligible for a Regents scholarship because I wasn't a citizen. My father had not taken out a citizenship application for me.

But I was eligible to take the Pulitzer scholarship exam, which was the College Boards. I wanted one of these Pulitzer scholarships. Apparently I did very well, but they never told me whether I had won until September. It was too late to enroll in any college. They told me I was among the finalists, but I just missed out.

CHAPTER 3

College Days—Brooklyn College and Cooper Union, 1936–1942

Starts Brooklyn College

In the fall of 1936, when I found out I wasn't going to get a Pulitzer scholarship, I decided to go to Brooklyn College to major in math. That's where my college days started. I was given a curriculum, and they told me what classes I had to take.

I told them I was going to major in math, so they enrolled me in math, English, chemistry and other subjects, not physics. I decided to go out for the soccer team. The schedule interfered with my soccer schedule, so it had to be changed. I still ended up with chemistry, English and math, and I think economics and German. So I had to take those five courses. It was a pretty heavy load. I did make the soccer team.

In the spring of 1937, I continued some of these studies, and I started taking physics and integral calculus. My first term in college I took algebraic geometry. I skipped differential calculus the first term because I took an exam and they decided I was good enough, but they insisted that I take integral calculus my second term. I had a very fine teacher named Rogers who thought I was one of the best students he ever had.

I started studying physics. The physics teacher wanted me to major in physics because I had done exceptionally well. In fact, at the end of the term they gave an exam, which he said nobody ever finished. I did. I think I got a hundred. He wanted me to become a physics major, and I could take advanced math courses.

Makes soccer team

Although I never played soccer here in America, I had learned in Europe, and I could run like a deer and I had a very strong kick. So I became a substitute because I didn't have the exact skills, but I could make a good wing. We used to practice with the first team.

I was on the second team. We'd practice there. We'd learn various things. They improved my kicking, showed me how to kick properly. We used to have these matches between the first team and the second team. I was maybe 135 pounds, very wiry. The captain of the team was maybe five feet nine, 160 pounds. One day we collided and he went down like a rock. I was really solid. He didn't get hurt, but I knocked him down when I tackled him to get to the ball.

We used to play teams on Saturday, and the star of our team was the son of a Rabbi who couldn't play on Saturday. I wasn't supposed to play because my father was religious, but I never told him where I went on Saturday afternoons. I had my gear stowed with my friend, Joe Weissboro, who lived a few blocks from me. I'd pick it up Saturday and go out to the soccer field to be ready to play.

One Saturday ... usually we played Friday but this time we played Saturday ... he didn't show up, so I was substituted for him, and I played halfback instead of wing, which was my normal position. I think we won the game and I almost kicked a goal. At the end of the term, because I played in a regular game, I got the letter, you know, the sweater with the letter "B" for Brooklyn College.

First taste of Chinese food

The most interesting part of this was the coach, because we had a modestly successful season, decided to treat us to a Chinese lunch. I'd never been to a Chinese restaurant. In fact, I never ate out because we had kosher food at home. And here we went to the Chinese restaurant and they served chicken chow mien. I'd never seen it before. I looked at it. It looked awful. I almost threw up. I couldn't eat it. Later on, when I got older and we came to Boston, I learned to like Chinese food. But that was my first experience.

Excels on math team

That spring there was an intercollegiate citywide math competition that involved both calculus and algebra. Brooklyn College had to form a team. My professor insisted that I be on the team. The head of the department objected because he said there were other kids who were majors in math, and they should be on the team. But my professor insisted that I was much better than they were. The department head didn't like it, but I was made a member of the team.

We took the exams. The team won first prize. I turned out to be the best player on the team, to the chagrin of the head of the department. In fact, I tied with Harry Schwartz [3.1] from Columbia, and they had to have a playoff to see who would win, and I beat Harry. So I came out number one in the competition, and I was going to get a medal for being the top scorer.

That spring we were celebrating at the math club. We were celebrating our victory. The head of the department was supposed to give me my medal. He refused, because I guess he was peeved that a freshman should beat out his prized students. So my professor pinned the medal on me. You can see that even teachers were very petty.

Runs afoul of communist professors

There were many communists both among the students and the faculty in Brooklyn College. This was downtown Brooklyn. There was apparently a march or something, a meeting objecting to whatever it was they were objecting to. I guess I was apolitical. I didn't bother with any of this stuff. I was too busy trying to get an education and playing soccer.

I was on my way to one of my math classes, and my philosophy professor came along and said, "Why aren't you at this gathering?" I said, "I've got a math class I have to attend. I don't skip classes." So he got very angry (apparently he was one of the communists). He said, "I thought you were one of my best students. You should have been attending this." This was for whatever the cause was, which couldn't interest me. I left him and went to my class.

Apparently the word got around, and all of a sudden in three of my classes, where I knew I was an excellent student, I ended up with a B or a C. I got a C in philosophy and he said I was one of his best students, which I was. That kind of soured me on Brooklyn College. But aside from that, I had very little money to continue.

Aces entrance exam for Cooper Union

So that summer, 1937, one of my friends from Boys High School, Stan Manson [2.9], who was a year behind me, was going to take the exams for Cooper Union. He told me there were scholarships available if you did well on the exams. So I took the exam that summer, and sent in my high school record after having finished one year in college. I took the exam and apparently I scored very well. In fact, not only very well, I was told upon graduation that I had scored the highest of any student that ever took that exam.

The exam consisted primarily of mathematics, visualization which apparently means seeing boxes and things. Apparently I was innately good at it. And of course verbal, which wasn't my strength, but I wasn't bad at it by the time I reached college age. My vocabulary had improved. The total score was higher than anyone had ever scored. The reason I found out, I was an honor student and I graduated Cooper Union, and when I was given the award, the president of the school announced, "I remember you. You were the boy who scored the highest of anybody who ever took the exam." I didn't know that.

Both Stan Manson and I apparently scored very high, and we both got what's called a Schweinberg Scholarship from Cooper Union. A thousand people took the exam, 100 were admitted, and among the 100, I guess he and I were the two top scorers, and we got, in addition, $300 a year for expenses. You needed it for books, clothes, and everything else. That made the difference whether I continued college or not. That's the reason I took the exam.

I couldn't have completed Brooklyn College anyway because I didn't have money. I didn't have money for clothes, carfare, and books. In fact, I owed my sister – who was working – money for my Brooklyn College books that I borrowed from her. So when I got the scholarship and started getting money every month, I paid her back.

That was a boon. It wasn't my intention to be an engineer, but that's what they offered at Cooper Union: mechanical engineering, electrical engineering, chemical engineering, and civil engineering. Those were the choices. I don't know what made me and Stan Manson [3.2] decide. We decided to become mechanical engineers. Electrical engineering would have been a better choice, but at that time I didn't know the difference.

So that's how I got into Cooper Union. It was quite a different school from Brooklyn College.

Enjoys first two years at Cooper Union

I enjoyed Cooper Union the first two years because there were general subjects, physics, chemistry, mathematics. I had to take math and calculus over the first two years, and they were general subjects. The last two years they were specialized in mechanical engineering, which I didn't enjoy. But I did what I had to do.

One of the courses I did enjoy, and apparently I did very well at, was mechanical drawing, and there was a related subject to it. So the first two years were apparently enjoyable. In the second year I had the highest marks of anyone, including Stan Manson, who ultimately graduated with the highest scores ever.

While at Cooper Union, I did engage in studying math by myself during the summers. So whenever I took a course, even on differential equations, I already knew it. Before my junior year, during that summer, I studied differential equations on my own so that when I took the course I knew everything and I was very good at it. I took this book by Cartwright. Did every problem in the back. So when they gave me the exams, I would finish in about 10 or 15 minutes and walk out. I ended up getting 100 in every exam.

At the end of the term at Cooper Union they gave you both a numerical mark and a letter. Of course I got an A and I got a 99. So I went to my teacher, who was Mr. Reddick, the head of the Math Department. I asked him, "Why didn't you give me 100? I got 100 in every exam." He said, "No student is perfect." I didn't give him a rebuttal.

Organizes math team

During the time I was at Cooper Union, in the first year, in the spring of 1938, I organized the math team to compete in the competition that I had won at Brooklyn College, the citywide math competition. Lo and behold, my team won the competition! At that time I didn't come out first, but I did quite well. I was the star of the team. The rest of the team consisted of Stan Manson and Ted Berlin. We did very well.

We competed in these competitions. The next year we also won the competition. We also competed in a nationwide competition which included Canada as well. I think it was called the Pi Mu Epsilon [*Pi Mu Epsilon is the US Honorary National Mathematics Society, founded in 1914*]. My team, which I trained, where I was both captain and coach, participated and came out third. Of course we were competing with math majors.

The next year, 1939, my third year at Cooper Union, I also competed in a citywide math competition and won it again. This time I met Melvin Lax [3.2]. The two Lax's came out first and second. I came out first. Melvin Lax, who was also a very bright fellow, came out second. This was very interesting. The two Lax's were supposedly the best mathematicians in the city. He was at NYU. I was at Cooper Union.

Takes honors course in math

When I was a senior, we were supposed to take an honors course. They wouldn't let me take it in mechanical engineering. So I took it in math. I decided I would take an honors course under the tutelage of Drs. Miller and Reddick, both of whom wrote the advanced calculus book [3.3] that they used. I studied the book and wrote a term report on Bessel functions and their use in solving certain kinds of problems in engineering. I graduated with honors in math.

Last two years are boring

But in my last two years I lost interest. I did just enough to keep my scholarship, but some of the courses, on steam tables and such, were very boring and mundane. But I managed to hold my own.

At any rate, Cooper Union was a great benefit to me. It did teach me one thing. They worked you very hard and gave you an awful lot of homework. So you acquired the discipline to do so. The engineering subjects were not easy. They involved a lot of studying and discipline. So I think that was the greatest benefit that Cooper Union gave me.

Unfortunately, by the end of the third year I had finished all the math courses they would offer in mechanical engineering. Advanced calculus was offered to the electrical engineers but not to the mechanical engineers, so I missed out on that. That was my big disappointment.

Electrical engineering was more mathematical, and actually more interesting than the mechanical engineering. But I didn't know that. There was a lot of lab work involved with steam engines and trips to power plants and things like that. They were very repetitive, like the power plant. Once you saw one, you saw them all.

The most interesting thing we visited was the Rupert's plant that made beer. In conjunction with that, there was a power plant. So we visited both the beer-making part and the power plant. And we visited many of the power plants around New York. Then of course you had to write reports. There were always reports, and those were not my favorite task.

You had mechanical drawing all throughout the four years. And you had to design an engine. That was not an easy project. I was a very careful and meticulous artist, and it took me a lot longer than most of the kids. I did a good job, but it took an awful lot of time. So when you went to school almost from morning to night, you were busy studying, and over the weekends too. Maybe I took off Saturday and Sunday mornings to play handball in a nearby court. That was my only recreation.

I remember Saturday night. On Saturday I couldn't draw at home, but Saturday night I was at the drawing board. Sunday I was busy with the studies, homework and drawing. So you were busy. I would say we went five hours to classes. We probably had at least 30 hours of homework. We started school at ten o'clock, and you'd get home maybe four o'clock. I'd start on my homework. I'd have dinner. From six o'clock on 'til about ten o'clock I would be working. I would spend about five hours a day on homework. Friday nights I usually took off and went to a movie. But on Saturday night and Sunday, I spent most of those days doing homework. I spent a good 30 hours doing homework. I would study on the subway, but that took half an hour. I would put in a good 55 to 60 hours some weeks. That was a lot of work. I would spend maybe Saturday morning and Sunday morning getting some exercise.

Moves to Manhattan after Brooklyn College

After I finished Brooklyn College, we moved to Manhattan. My father got another job, a much better job as a sexton. Our first apartment was on Broadway, and you could hear the trucks at night going and coming because our apartment faced the street. But I guess I managed to sleep. We lived there all the time that I went to Cooper Union. I would take the subway there. That was a very nice area.

We lived right at the foot of Fort Tryon Park, which is a park that the Rockefellers donated to the city. At that time you could walk to it without worrying. During the fall and the spring, when I had to do homework and it was still daylight, I would take my books, sit on a park bench looking over the Hudson and study, which was very nice. It was always fun walking around Fort Tryon Park. They had high bars there and a basketball court up on the top which I could use. When it got colder and you couldn't play handball, I would wear gloves, and go up there for my exercise Saturday and Sunday mornings. It was a very pleasant place to live.

During the summers I joined the Miramar Club, which was about a mile away, which had a swimming pool and handball courts. I would spend my time there, but I would work part time.

WPA job and other jobs

While I went to Cooper Union for the first two years, I had a WPA [*Works Progress Administration*] job. The last two years, I lost it [WPA job] because I wasn't a citizen. I don't know why in the first two years I was allowed to work.

One of the jobs I had during my Cooper Union years, during the summer, I think when I got to be a junior, at the end of my junior year ... they let me mark the papers, both in the Arts School and the Engineering School. So they had a professor and myself and some of their own faculty during that summer marking the exams, the same exam that I took, in order to admit students. Apparently, they felt I was bright enough and trustworthy to mark both the art and the other exams. That occupied me for maybe two weeks or a month.

Routine while at Cooper Union

The second year, when I scored the highest, I was working on Saturdays, and sometimes after school for a few hours, at Cooper Union in the library, putting books back, cleaning the dust from the bookshelves, and various other things. Sometimes handing out books. Cooper Union had a public library that was attended by many of the homeless that used to sit in the streets. In the winter they would come up and sit in the library and they would borrow books. We'd let them have them. They couldn't take them out of the library, but they would read them there. That's the way I made extra money. I also used to tutor some of the other students. I also sang in a choir. So in addition to the tuition, I made extra money which I used to give to my mother. That's the way I spent my college years. Then I graduated in 1941.

Highest entrance exam marks

At graduation, in 1941, all the people who got honors with their degrees were called to the podium. There the president introduced me and said, "I know you. You're the student who got the highest marks ever in the entrance exams to Cooper Union." That was the first time I realized why I got the Schweinberg Scholarship. Apparently, I had scored very high in the math and the visualization, because I know in the verbal I was better than average, but I wasn't outstanding. So I must have scored extremely high in the other two parts of the exam.

Discrimination in job interviews

I graduated, and we had interviews. But none of the big companies interviewed the Jewish boys. The four top students, including myself, Stan Manson, and two other boys, all Jewish, would not be interviewed. All four of us were good friends. One was Joe Heiman, who was half-Jewish–half-Christian, and he learned the system and said he was Christian because his mother was Catholic. So he got an interview from RCA, but the other three did not. In spite of the fact that Sarnoff, who was the head of RCA, was Jewish.

We did not get interviewed by GE. There were other students that had lower scores than we did who did. I did have one interview from a firm here in Massachusetts. It went very well. At the end he asked me what was my religion. I told him it was Jewish. I never heard from them.

Accepts job at Curtiss-Wright in Buffalo

So there was discrimination. But I did get an interview and a job offer from Curtiss-Wright. I accepted, but that was in Buffalo. I actually wanted to stay in New York City, but none of the outfits, Bell Telephone, all the others, would interview us. So there was discrimination. Joe Heiman went to RCA. Stan Manson got an offer from United Technologies. I don't know why I didn't.

Anyway, I got a job with Curtiss-Wright. I was very pleased, but I had wanted to live in New York City.

So I went to Buffalo. I rented a room from a Hungarian couple. There was another fellow from New Jersey, also a graduate of engineering, I think from Stevens. We each rented a room from this couple, who didn't have children. The wife was delighted to have me there. She remarked that I was so neat. The other guy wasn't so neat. He was a big husky fellow. He had played football. He and I used to wrestle on the beaches at Lake Erie, and I don't know how I managed. But we got to be good friends. We spent the summer being trained by Curtiss-Wright in the techniques of reading *blueprints* and the various aspects of the aircraft industry. I intended to go to night school but there were no night schools in Buffalo.

Takes Army job in New York City, night courses in math

I decided I had to continue my education, so I got a Civil Service job with the US Army Corps of Engineers in New York City and left Curtiss-Wright. I came back to New York City and decided to go to night school at Brooklyn Polytech to study applied math and related subjects. I took three courses, two courses at Brooklyn Polytech and one course at NYU.

The job with the Corps of Engineers was to oversee the operation and the maintenance of tugs and barges that they used in maintaining the harbor, dredging it and all the other functions. I didn't enjoy that job too much, but, anyway, I went to night school.

That year, 1941, was a busy year. I had some interesting times on my job. I used to go to Staten Island to oversee the repair of barges at some shipyard. I would go on the ferry. Occasionally, we would go out on a fast boat, a confiscated smuggler's speedboat. We would inspect the tugs and the other boats.

One day it was very rough in the harbor. My boss and I were going to visit the tugs. The little boat was really bouncing up and down. I was starting to get seasick. Then we went down below in a tug which was very, let's say, dank, and the air was foul. My boss said, "Ben, you look green." I was ready to throw up. We got on top of the tug and we transferred to a bigger tug which was much more stable. I had some fresh air. I recovered. That was the closest I came to getting seasick since the time that we came to the United States on the *Olympic*.

Meets Blossom, takes too many courses

At the same time, I met Blossom through a friend of mine from Cooper Union, Ken Robinson. I fell in love with her.

Well, that was an overload. I got very little sleep and I lost a lot of weight. On Pearl Harbor Day, December 7, 1941, I was in a hospital with a very serious case of gingivitis and high fever. Apparently, I had not taken proper care of myself. I just wasn't getting enough sleep.

So I decided to cut back on my night courses. I had this job with the US Army Corps of Engineers, and I had medication, so in the spring I took it easy. I took just two courses, including another course in advanced calculus, at Brooklyn Polytech.

Fails, then passes physical exam for the draft

In the spring of 1942, I had to be tested for the draft, so they sent me to a doctor. As a consequence of my lack of sleep and other things, I had a bad case of psoriasis. My knees, my elbows, even my face broke out. So when I went for my physical exam, the doctor classified me 2B or whatever it is. I wasn't eligible for the draft. I was very upset about it. During that time I met a doctor at the hospital where I was treated, which was right near our house, in fact, within walking distance. He told me I had psoriasis which I had not treated properly. He started treating me. With more rest and less anxiety, the psoriasis went away. In other words, it was cured.

So I went back to the Draft Board and got classified 1A. I was proud. That wasn't very smart of me, but it's the way I felt. I felt I was a healthy strong specimen.

Starts graduate school at Brown University in 1942

I decided I wanted to go to graduate school. I went to my Draft Board, told them my plans, that I had gotten a summer fellowship for Brown, and I also had a fellowship waiting for me at Illinois Institute of Technology (I had applied there). I would go there for a year and get a master's degree and then work in the defense industry, probably at the NACA [*National Advisory Committee for Aeronautics, which would become NASA in 1958*]. Stan Manson, who already was getting a master's degree that following spring of 1942, had gotten a job in NACA in Cleveland, and that's where I wanted to go. I thought they had agreed.

So I saved some money, went to Brown, got a room, and got a part-time job as a draftsman with an engineer who spent most of the winter in Columbia. In fact, he liked my drafting so much he wanted to hire me to go back to Columbia with him to join his staff there (he was an American engineer), which of course I didn't want to do.

I went to Brown and started studying mathematics. There I met some interesting people, including Luttinger [3.4]. We later came to know Luttinger. He was 15 years old and he was taking the same courses in advanced math that I was taking [3.5]. This was vector analysis, differential equations and, I think, stress analysis, which was a mathematical course, which I took because I intended to be an applied mathematician. I did so well they offered me a full-time scholarship for the fall.

But at the beginning of August, the Draft Board informed me that I was to be drafted. So I got on the train and was on my way home to join the Army when Dean Richardson, the man who gave me the scholarship, said, "What are you doing here?" I said, "I've been drafted." He said, "No, we are going to appeal."

I was supposed to report in two weeks. They sent through the papers and I was to have another hearing, so my report date to the Draft Board was delayed. So I went back to Brown, continued my studies. But about the middle of August, I was informed by the Draft Board that, no deal, I had to go into the Army.

Elopes with Blossom, drafted into the Army

In February 1942, Blossom and I eloped and got married. I lived in my folk's house and contributed to the maintenance of the home. Blossom, who was working as a secretary, lived with her mother.

Then I got my draft notice. Eventually, after an appeal, I was finally drafted. But I did manage to get an award of a fellowship to Brown, with which, if I had stayed, I could have gotten a master's degree in applied mathematics. But I had to go into the Army.

Once I got my draft notice, Blossom and I told our parents we were married. And we decided to go on a honeymoon. So the last two weeks in August 1942 we went on a honeymoon, to the Catskills.

CHAPTER 4

Army Days and the MIT Radiation Laboratory, 1942–1945

Enters the Army

In August 1942, on the date I was supposed to go, I reported to the Draft Board. They took us to Pennsylvania Station, and then we went by train out to Camp Upton, which is on Long Island. It's exactly the location where Brookhaven National Laboratory is now.

We had physical exams. They looked me over and decided I was fit for the Army. They gave you all the inoculations, just about everything, tetanus shot and all the other shots, just in case you went overseas or something. Then we had lunch, and I guess some other introductory session. We got our clothes with a big duffel bag. These were our fatigues and regular clothes with our shoes, hats, everything. Then we had dinner. It was a tiring day, a full day. Right after dinner, we were to take our duffel bags and go to the other side of Camp Upton to take some tests.

There we were, dead tired, sitting, taking these tests, which were very similar to, I guess, IQ tests. I didn't know that, but they had the verbal, mathematical, visualizations. Not dissimilar, but maybe a little more elementary than the sort of things I took in Cooper Union. As tired as I was, I was a good test taker and I think I did pretty well. I didn't know until later how well I did.

The next day we were assigned to barracks. Saturday and Sunday I was assigned to kitchen duty. There I learned to use one of the machines to peel potatoes, so I didn't have to peel by hand. I guess I was one of those people who adapts very quickly.

Blossom visits

Sunday afternoon we were supposed to have off. I was in my fatigues. I was told my wife is waiting for me. So I left the kitchen, or wherever it was, and went to the PX, or the enlisted men's area, where we were allowed to see visitors. Blossom saw me in my fatigues and she burst into tears. The suit was a little loose on me, and the cap was like, she says, "You look like Dumbo." And I laughed.

We were happy to see each other. I guess we went into the enlisted men's area or the PX, had some drinks. Then she left for home and I went back to the barracks.

Basic training at Fort Monmouth

Soon thereafter we were assigned to basic training at Fort Monmouth, New Jersey. We went from Camp Upton by train to Fort Monmouth. There I received a real education, and saw a broad spectrum of what kind of people come from all over, and what kind of people we have in this country. Some of it was favorable and some of it was not.

Most of the people were relatively crude and uneducated. Just about 90 percent used the "f" word for everything to express themselves. Particularly in my barracks. I shared the barracks with an American Chinese, born in America, who could barely speak English. He came from San Francisco. Apparently he was brought up in Chinese school, but since he was an American citizen he had to be drafted.

Another was an Irish lawyer who was a fairly intelligent man. Another one was an auto mechanic, a young fellow who had just recently got married to a fairly pretty girl. He couldn't talk and think about anything but getting back and jumping into bed with her. That's all he talked about. The last one was a real character, he was a rigger. He had married an older woman with a son. He hated them both. Every other word was the "f" word. This was a really crude, uneducated man, but apparently an expert rigger. I'm sure he loved his wife, but he would curse her and talk about her in very derogatory terms. So here was a cosmos of what America is like.

Takes more exams

So we got basic training. We went to the rifle range. We also were given a handbook in which we were supposed to learn all the things you're supposed to know about the Army. At every chance, I would read and study the handbook. Having a good memory, I could commit it to memory.

Eventually, we were given an exam on that. Apparently I scored very high, and everybody knew that I had scored high. In fact, some of the other people from some of the other barracks would say, "Gee, you're a brain, but otherwise you're not a macho guy." Or words to that effect. I was a little skinny guy. They used different terms. But, in other words, they looked down on your intellectuality. But the mechanic scored very high on the rifle. He was a hero. So the lawyer said, "Well this is typical of the Army. They admire physical prowess, but intellectual talent doesn't count here." In other words, it wasn't respected, and this was the case.

Assigned to radio school at Fort Monmouth

After the basic training, I guess because of my scoring on the exam, I got a nice assignment to the radio school right at Fort Monmouth. Which meant I was going to be a radio mechanic. It was self-paced course. So I took to it. I never had electronics, but I found it fascinating, even though it was fairly elementary. Of course, I was well-versed in electricity and magnetism from my Cooper Union days, but not in electronics. I went through the course in about three weeks.

Assigned to Officer Candidate School (OCS)

Now, I had heard of Officer Candidate School (OCS). So, when I was about ready to be getting an assignment, I went up to the captain. I forget the captain's name, but he was a very intelligent man. I said, "I'm a very good mathematician. I could be much more useful to the Army here at Fort Monmouth teaching mathematics courses for OCS." So he says, "Let me think about it." The next day he called me into the office, and he says "You can't teach unless you become an officer. I looked up your IQ. Your IQ is 154 based on that exam. You and another man who has a similar IQ are going to OCS."

Isn't that amazing? This was a very intelligent man. He didn't single me out. There were two of us who scored very high on the IQ exam, over 150, and we both went to OCS. We went to different barracks. There began my days at OCS. So, exactly six weeks after being inducted in the middle of August, there I was in OCS. Immediately, I automatically became a corporal. That was a promotion.

Fellow students in OCS

There we had a somewhat more intelligent group of people, or people a little bit more talented in different ways. We had one guy who was a professor at CCNY. He was a very bright man. A couple of lawyers, some businessmen. There were a couple of people from the entertainment field, one of whom worked for the man who was the chief assistant to *What's My Line,* the program where they interview people. There was a young fellow who was regular in the Army who really knew the stuff. He was really a beautiful specimen. There was a regular soldier, a sergeant, who had difficulty with the courses, but certainly knew his Army techniques.

So there was a broad spectrum, but I think a much more, let's say, intelligent group than the ones in basic training. I had met southerners and others who respected shooting more than intellectual ability. Here it was different.

We had courses in arithmetic. The sergeant had a difficult time. He had to take it three times before he passed that course, so this was his third class. He finally did pass. There was a course in electricity and magnetism, Army bookkeeping or whatever it was, and a few other courses. This was OCS for the Signal Corps. Certainly in the math course and the others I did extremely well. I probably was the top student there.

Non-kosher food in basic training and OCS

When I got into the Army, the food wasn't kosher. My father said, "That's fine, this is an emergency, you're fighting for your country. The Jewish law allows you to eat non-kosher food."

Both in OCS and basic training, we'd get up at five o'clock in the morning, run around for half an hour. All of us were so hungry. We were young, in our early twenties. You'd smell the food when you got into the chow line. I used to smell the bacon. The bacon tastes and smells like something my mother used to make. My mother used to cook chicken, take the chicken skin and render it in the fat. It's called griven. Griven and bacon taste and smell alike.

I was so hungry, so I decided to taste it. I loved it. I didn't tell my father that I was eating bacon. He said you could eat non-kosher food, so what difference did it make? This was the first time I really started eating non-kosher food on a regular basis. I did it while I was at Fort Monmouth, both in basic training and OCS.

Bright guy, small size

At the end of the first month, we were supposed to rate each of the people. They rated me very high. The comments by most of the people were, "Ben Lax is a brain, but he is not going to be a leader." How wrong they were! But that's the way it was, that I wasn't going to be a leader.

Because I was one of the smallest people in OCS, and there were other people like me, the tactical officers went out of their way to try and break us. Apparently in the Army size and physical prowess counted, even here in the Signal Corps. They went after me mercilessly. They would pick up my mattress and they'd find a thread underneath, and I would get a gig. I would get a gig for the least things. They were trying to break me.

They decided to hold an obstacle course. The obstacle course consisted of an eight foot wall that you had to jump up and climb.

Now, I was very spry, and I was a gymnast. At that time I had lost some weight. And I was particularly strong. I jumped on the wall and just practically almost pole vaulted over it. I pulled myself up and over, and I ran like a deer. I came out number one. After that they stopped picking on me. Apparently, physical ability counted more. In fact they made me a little squadron commander. And I didn't know any of the Army tactics.

Assigned to radar school at Harvard and MIT

A young fellow in one of the other barracks informed me there was a school at Harvard and MIT that taught radar. So I went up to the tactical officer and I told him that's where I wanted to be assigned. This was about two or three weeks before graduation, after I had exonerated myself. Sure enough, when I got my papers after I came back from home that weekend, I was assigned to Harvard and MIT.

So, by speaking up and asking, apparently the Army did right by me. And by asserting and demonstrating my intellectual capabilities, they had the good sense to send me to Harvard and MIT. I thought that was great. So that's what happened, and that's how I got to Cambridge.

Buys officer uniforms

Then I got my orders to go to Harvard and MIT, the radar school, which is what I asked for, and that was great. In between I had a week of vacation in which I was supposed to get officer's uniforms. So I went home to New York.

Blossom was at her mother's house, and I stayed at my father and mother's house. We spent that week shopping for my officer's clothes. They were very fancy. These were the winter clothes because this was in January of 1943. I got a fancy dark coat, light-colored pants, and the whole outfit, new shoes, the whole bit, hat. I looked like an officer.

Takes train to Cambridge

At the end of the week, about the middle of January, vacation was over. Blossom and I got on the train, left in the afternoon, and went to Boston.

FIGURE 4.1   First Lieutenant Benjamin Lax, *circa* 1944. (Photo from Daniel R. Lax.)

We got there. We went to Harvard Square, and we started making calls trying to get rooms for the night. We had lots of trouble, and it started snowing.

Finally, after supper, we got a room across from Harvard Yard, on Broadway or somewhere there, in a private home. There we were with our baggage. By this time, the snow was pretty deep. We walked across Harvard Yard with our bags. We didn't have that much. We were young, and traveled lightly anyway. I had my officer's topcoat, which was very good, a nice warm coat. We probably got there about nine or ten o'clock. We got this room, and we got into our night clothes and went to bed.

In the middle of the night, Blossom woke me up. "I hear noises." Sure enough, there were rustling noises. It turned out there were mice in the room. They were running around. She got kind of scared. I quieted her down, and we went back to sleep. We got up early in the morning the next day. We went back to Harvard Square, to one of the local restaurants there. Had a nice breakfast, and immediately went to the public phone and started making telephone calls.

Small apartment in Brighton

I must have gotten the newspaper, because eventually I called Washington Apartments in Brighton. It was a lady, a Mrs. Rich, who answered the phone. She said, if you come right away, we've got a little apartment for you and your wife. Immediately we grabbed a taxi, got there and got our apartment.

It was an apartment on the first floor. It had a kitchen, a foyer and a living room, all furnished. There was a couch, which you could pull out to go to sleep. That was our bed. We got settled. There were a couple of closets. So, Blossom and I started our married life essentially in this little apartment in Brighton, on Washington Street. That became our home for quite a few years.

Begins radar school at Harvard and MIT

The next day I went to Harvard, to the radar school. This radar school accommodated all the services. A colonel [4.1] was in charge. He was an ex-astronomer from Chicago. I reported to him. I was told that I was going to go for three months to study electronics, basic electronics, which I never had.

Some of the people did have it. This included all sorts of people, all of whom were from the Navy, the Army, and the Air Corps. There was one man from the Marines. We started taking these courses.

The courses were given by some of the professors, one of whom was Le Corbeiller [4.2] from France, who was a fantastic artist. Another was Professor King [4.3] who used to teach electromagnetic theory, and Mimno [4.4] and others who taught basic electronics. There was one who was a specialist in vacuum tubes. We learned about vacuum tubes. We learned about simple circuits, resonant circuits. Le Corbeiller drew beautiful diagrams of resonance and impedances. So essentially, we learned fundamental basic electronics and electrical engineering related to vacuum tube electronics, which was the thing everybody used.

We'd go to classes. We'd get there, I don't know, eight or nine o'clock, and then we'd be there all afternoon. It was a very intensive course. We'd have laboratory, three hours of labs, and lectures. At the end of the week we'd get tests.

Because electronics was easy for me, I did well. I just barely missed the top 10, a disappointment for me. But Murray Winnick [4.5], who was an electrical engineer and an electronics expert, got among the first ten, and we got to be good friends. That's when I met him.

Continues radar school at Harbor Building

Then, when we graduated, we were assigned to the radar school, which was downtown in the Harbor Building.

There we started learning about radar. The MIT professors were the teachers, and Bill Radford [4.6] was one of them. We took courses from them. And there were some men in uniform who also taught courses. It was mostly MIT professors. They taught us all about radar. We essentially learned about the L-band radar, which was prevalent. And we got a little introduction to microwaves, but barely.

It turned out again that we studied a lot. I used to stay at night and study because the lecture notes and manuals were classified so you couldn't take the stuff home. So I would stay there and I would study evenings. I would get home a little late.

I ended up as the number one student, even though the first time I didn't make it. But now that I had the basic electronics and I was studying hard, I got the highest marks of everybody.

Assigned to MIT Radiation Laboratory

The colonel [4.1] who was in charge of both the Harvard and MIT programs decided that I would be assigned to MIT Radiation Laboratory [4.7]. So my next orders read that I would be a radar officer at the MIT Radiation Laboratory.

But there was a glitch. It turned out, because I was born in Hungary, the Air Corps or another of the services decided (actually, my headquarters was at Fort Monmouth) that I must be cleared, once more, to make sure I'm not a spy.

Becomes interim instructor at Harvard Radio Research Laboratory

So, the colonel decided to make me an instructor in the lab at Harvard while I was awaiting clearance. So at the beginning of July, I started working there. This went on until the March of next year, 1944. It took eight months to get me cleared.

Both Murray Winnick and I were working at Harvard, and we got to know each other. I guess I was there longer. And what happened was that I got to learn. I made circuits and I got to learn more. I had spare time. I was practically a free agent.

After I did my duties, I would go to the Harvard library and start studying math. I sat in on the course given by Van Vleck [4.8] on quantum mechanics, which I didn't understand because I didn't have the proper background in atomic physics or in physics. I also sat in on King's [4.3] course, which was given by one of his colleagues, on electromagnetic theory. This I could understand. I got the book on electromagnetic theory and I learned about waveguides and microwaves. So, I didn't waste my time while I was there.

Colonel Fox leads Saturday morning exercises

Philip Fox [4.1]. That was the name of the colonel in charge. He was a wonderful man. He was an older man. He used to lead us in exercises. This was in the spring of 1943, before I became an instructor. He would get into a position to do push-ups. He would push himself up, clap his hands, put his hands down. I think he was in his sixties. He was not only a bright man, he was quite an athletic figure of a man.

He would do his antics for us on Saturday mornings at Harvard. Even if we were at MIT on Monday through Fridays, we also had to appear at Harvard for our Saturday exercises.

Occasionally we would be asked to go and learn how to handle a pistol. That was the extent of our military training.

One day I missed the trolley, and I started running for it and caught it, because, you know, as I told you, I used to be a soccer player. Murray Winnick said, "I knew it was you. You're the only guy who could do that." We got to be good friends. We both were assigned to the Radiation Lab, but I lost track of him. He was in an entirely different division.

Li'l Abner radar

There I was, in Division 10. Professor Ernie Pollard [4.9] from Yale was the head of it. Jack Millman [4.10] was the group leader. He was a Professor of Electrical Engineering from Columbia, and nominally he was my boss. When I got there, they were planning a new radar, which was conceived by Ridenour [4.11], called Li'l Abner [4.12].

This radar was supposed to be man-portable. The concept was that every component of it was able to be carried on the back of a GI, each of the pieces. [*That's how this radar got the name Li'l Abner. The GI would have to be as strong as the legendary Li'l Abner of comic strip fame to be able to carry any of the pieces up a mountain.*] They could carry it up a mountain. It would be a height finder. It was to be an X-band radar. It would have a vertical antenna that would scan up and down and determine the altitude.

It would be mounted on a circular rotating track, a machine gun track, which was about five feet in diameter. All the components would be mounted on that. You would rotate that by hand. But the original concept had been modified. I must admit this was a very stupid decision, which ultimately I reversed. It had been decided that the RF head would be mounted on the antenna itself, and would nutate with the antenna. I think this was a very bad idea.

The components were being built, and I undertook the job of assembling the breadboard model, which means we got an antenna. We got an RF head, which we put up on the top. We had a receiver. The RF head and the receiver were on the top, including the magnetron. We had cables from the power supply, which was a small putt-putt motor which provided the power for that and the motor that ran the antenna. And we had an X-band waveguide and a horn.

We built up the infrastructure with four-by-fours and mounted the antenna on the track, and just rotated the arm in the azimuthal direction. We assembled it down at Hanscom Air Force Base in a little shed.

I had four GIs who were working for me. We got all the components. I worked very closely with the components group in the laboratory, including the test equipment group, the RF group. The modulator was down below, and it went up on cables in a sort of homemade fashion.

We started tests. We had flight tests. This was done in the summer of 1944. So, the summer and early fall of 1944 we were testing this radar. I was running the tests with my four GIs.

At that time Sam Levine [4.13] came and was part of the group. He had come from Panama. He was in charge of a radar station there. He knew his electronics better than any of us, so he was sort of a consultant, but I did most of the direction.

Takes charge of project

I just simply took charge. Became *de facto* project leader. I integrated the components with the rest of the lab. The Group Leader, Belmont Farley [4.14], and Tom Moore [4.15], the assistant group leader, attended all the meetings, but really didn't understand the system of the radar. I knew every vacuum tube and circuit in that system. The only thing that bothered me was the RF head being on the antenna.

Tests breadboard radar in mountains of North Carolina

In the early summer of 1944, we took the breadboard radar down to Asheville, North Carolina in a truck. We wanted to test the breadboard radar in the mountains. It was supposed to be a mountain radar set.

We spent two weeks putting the components on a truck. I remember I worked so hard with my GIs, packing everything (it took two trucks), that, at the end of the two weeks (it was a hot summer), I lost about ten pounds. I was told, if I'm going to go down there, I better wear a pistol. I don't think I ever shot a pistol in my life!

In the early summer, I think it was the end of June, we packed these two trucks. Sam Levine was in one truck, I was in the other truck, and there were two GIs with each of us. Maybe we had an escort, an escort car. We started out from Cambridge. The Division decided to have an expediter help us on our trip.

He turned out to be a drunkard. I guess he was a business type whom they hired to help out in odds and ends. Every time we approached a city, before we got there we'd stop to eat or something, he would call the police, and we would get a police escort. We went through New Haven with a police escort and we got into New York State. We were going to go over the George Washington Bridge. Before we approached it, I think

FIGURE 4.2 The Li'l Abner radar set, designed mainly for mountainous country, saw action in the Pacific Theater late in the war, on Iwo Jima and on Okinawa. (Reproduced with permission from the book *Five Years at the Radiation Laboratory*, an internal publication of the Massachusetts Institute of Technology, 1946, page 170.)

in Connecticut, he called the New York Police. By the time we got to the Henry Hudson Parkway we had a police escort. This was rush hour, so they would go blazing fast. Then, in New Jersey, the scope fell out of the truck. The scope that we used for the height finder. Boy, we got scared. Gee whiskers, what the hell are we going to do if that doesn't work? We loaded it back on and tied it up. We said, from here on, no more escorts, particularly when we approach Washington. We got rid of this guy.

We drove through. The GIs did the driving. We got to Asheville, I think a couple of days later. We got to Asheville at night. We got up into the mountains. We assembled the radar set.

Sam Levine [4.13], who was a pretty good electronics man, found out that what happened was a couple of the tubes broke. We had lots of spares, so we replaced them. Everything was in working order. The tests were very successful. They had planes flying over and my GIs and I would run the operation. Sam Levine and I were in charge of the operation. It proved that all the components in this breadboard model worked as we expected.

It still had the antenna up on the back, and had this little putt-putt engine. But it worked. Things got more complicated after that, but this was the breadboard model. We just ran it the same way we ran it in the shed. So it worked! It was a very interesting time, and it was very exciting. I really enjoyed it.

Gains confidence in leadership abilities

I learned that I could take charge and direct other people. This was my first opportunity to do such a thing. It was amazing that I naturally took charge. I kept thinking of what my colleagues in the OCS said, that I didn't have leadership qualities. Among all of us that I met later on, I was the one who had the most leadership qualities. So, physical size isn't necessarily everything. It may give you the wrong impression.

By the fall of that year we had the radar tested. In the interim, the production system was being designed by one of the engineers in another group. He was designing it, and my job, which I assumed, was to connect all the systems, order all the cables for the production models, and all the components. Knowing all the circuits, I just went everywhere.

At one of the meetings with the Air Force, I pointed out that the power supply that supplied the whole system was inadequate, in that the modulator and the RF head and the motor took more power when you started the motor. It could go on dead center and it was dangerous. In fact, one of my GIs once took it off dead center and broke his arm.

I said, "This can't be. You need more juice." I said, there was another power supply that was bigger, and I chose the connector so that it could be compatible with that big power supply. But the major who represented Fort Monmouth headquarters objected: "It's got to be Li'l Abner. The power

FIGURE 4.3 First Lieutenant Benjamin Lax, *circa* 1944. (Photo from Daniel R. Lax.)

supply should be able to be carried on your back." But, I said, the radar wouldn't work properly.

I lost that one. In the meantime, we were designing the systems and I was coordinating all the purchases for the production line. In the early winter, I guess it started in November of 1944, I assembled the first prototype in the laboratory, tuned it up, and it worked. It worked like clockwork.

Successful field tests in Florida

In November, it was sent out to Florida, near Orlando, to be field tested, where they were going to have flights. It went out in perfect shape. Here at home we had lots of signals. Then, a week later, we got a call that they couldn't see anything on the radar. Tom Moore, the assistant group leader, went down there and fiddled around with it and screwed it up.

Immediately, the Air Force and Ernie Pollard decided that I was the man to go down there and fix it, because I had tuned it up and knew the whole system. So I went down there with orders for ten days to fix it. It took me one day to fix it. What Tom Moore had done was burn out the crystal detector, the "cat's whisker." Immediately I knew what was wrong. So I immediately replaced it. He also detuned the TR [*Transmittter/ Receiver*] and the anti-TR boxes, so I tuned them up. When I tuned up the anti-TR, we even got a much better signal than we had at home because I hadn't touched the anti-TR. I was told it was already adjusted, but since he had fooled around with it, I decided to tune it. We got signals, very strong signals, right off the bat. From there on, all I did was just come in and watch the radar.

It was working very well. Occasionally, when the Raytheon magnetron sparked and a crystal detector would go, I would test and replace the crystal detector with a new one. I had sent plenty of spare "cat's whisker" crystal detectors. The tests were very successful. They went on for ten weeks. I was supposed to have gone back in ten days, but the Air Force had a feud with Ernie Pollard and he was away in Europe. They decided Ernie Pollard wasn't going to dictate where one of their soldiers was to be assigned. I couldn't go home without orders.

So I and two other friends of mine who were helping with the tests ... they were officers ... rented a house. We were having a good time during the tests. We were eating oranges from the orange grove next door. One day we dug the "dead man" for the tower. While I was digging, there was a beautiful snake. I yelled, "What a beautiful snake there is here." And a guy said, "Get the hell out. That's a coral snake." It was at the end of my shovel. So I got out and he caught it. That was the last time I ever dug there.

In the evening we would go out and go bowling. My bowling score went from the low eighties to 160 while I was waiting to get back. At the end of ten weeks, DuBridge [4.16] and Ernie Pollard came to visit and Ernie Pollard yelled at me. He said, "Ben Lax! What the hell are you doing? The system [*back at the Radiation Lab*] is in a mess. You were supposed to take care of it." I said, "I can't leave Florida without orders." The next day I got orders. So I came back. He was right, all the components I ordered and everything was in a mess. Things weren't organized. My four GIs were not being supervised. I took over while the finishing touches to the design and construction of the components for the production line were made.

Production line at Logan Airport

I was in Florida in December and January. I came back in February 1945. By the end of March I had the whole thing reorganized, and we started a production line out in Logan Airport. There were something like a half-a-dozen garages next to one another.

I suggested that each radar be assembled by the squadron that was going to take it out into the field. The Air Force adapted that scheme. So these officers with their GIs would come and I would assign each of them. There were about a half a dozen based there. They would assemble, and I would make sure all the components were there.

They were assembling them. The only thing I wouldn't let them touch was the IF strip in the receiver. I was the only one who could tune it up because that was a delicate job. They were stagger-tuned, and that had to be done carefully. You had to know what you were doing. I got to be a pretty good electronics man by that time. Each radar set worked up to the standard that the prototype did. It would take them about two weeks to assemble it, run it, and learn to operate it.

I had some sort of instruction book that we wrote up for them to operate it, but mostly they learned by actually being trained on site. It was June 1945. By that time it was too late to go into the European theater, for which it was intended for the mountains. So it was shipped to the Japanese theater where it saw action.

Removes RF head from the antenna

While I was at Orlando, I decided that this idiotic business of putting the RF head on the antenna was stupid. It should have been put on the track. In between you could use rotating joints, which had been invented by some of the Radiation Lab people. Put that in there and leave the antenna without the RF head.

So, without consulting anybody, Captain Levine [4.13] and myself conspired. He got all the components and sent them to me while I was there during those ten weeks. We put it on and it worked like magic. I wrote back and called Ernie Pollard. I said, "This is the way it has to be done." So the production lines ultimately were designed to do it that way. We had accomplished something there.

Advocates for bigger power supply

I'll tell you the power supply story. I gave oral instructions to every cadre that, when they come into the field, they should throw the little power supply over the hill and requisition the other, bigger power supply. Several of them wrote me that they did just that. That was unorthodox, but I was not a military man. I was a very independent cuss. That's the way things worked out. Apparently, it was a very successful operation.

Pollard was a group leader, but I built the radar, not him. See, I actually directed the building of the radar, testing it, modifying it to improve it. Pollard said, "If Ben Lax says it's so, it's so." He gave me tremendous recommendations. That's why I had no trouble getting into Harvard and MIT. Of course, I had a good scholastic record, but he thought I was very talented. He and I didn't get along at first, but then he began to appreciate that I was a little fresh, I was brash, but I was competent. So he got to like me.

Apparently, I acquired a good reputation for my work. Sam Levine also contributed. He participated in the assembly line, giving instructions to the GIs. Between the two of us, we were the instructors. That's all it took. It was amazing what we could do with four GIs and two officers. We were essentially the people on the set, the people who designed the components. The group leaders had nothing to do with it. It was just the four GIs. There were very few civilians who were there in operations at Logan, except for those who supplied the components.

Experiences at the Radiation Laboratory

I met Jerry Heller [4.17] at the Radiation Lab. Jerry Heller was part of the test equipment component group. On the receiver end, the man I would talk to was Herb Weiss [4.18]. I met Pound [4.19]. When I wanted to know something about the magnetron, I would go to see Clogston [4.20] and Mel Herlin [4.21]. I got to know all these people. I would interact with these people.

The only one I didn't have anything to do with was Purcell [4.22]. I met Pound because I was interested in his balanced mixer, which we didn't use on our radar, but it was a very interesting microwave development.

The Li'l Abner project didn't have a high priority. Project Cadillac [4.23] had all the high priority. I was a little, skinny, boyish-looking lieutenant who would go around and make personal contacts with the purchasing department. They'd do anything for me.

I was a very enthusiastic guy. People were so taken by my enthusiasm, they would do anything for me and skip priorities, so I got everything I wanted by personal contact with the people, which I'm sure most group leaders didn't do. I got to know the purchasing people, and also the components people. I was essentially the immediate contact on all the purchasing. I ordered all the cables and connectors for the entire radar set. I would tell the components people what connectors and cables to put on to connect one system to another. I drew the schematic diagram for the entire radar. I would identify the components and show how each would connect to the other.

The Li'l Abner radar was separate from the regular PPI radar. The Li'l Abner radar would see a target. It would identify the altitude and the azimuth. It could stand alone.

Li'l Abner radar predicts weather

It was also a good radar set for weather. We discovered this one day when we were testing the breadboard. We saw a big "curtain," maybe about 30 miles away, on the scope. From the scope, we decided it was a squall. Our shed had a flap which we could pull up, a canvas tarpaulin, a curtain that goes on the front. So we immediately let it down. By the time we let it down, pretty soon the squall hit us. If we hadn't done that, the rain would have come in, and, because the radar set was pushed up against the front, the rain would have poured on it. We decided that it was a good weather radar.

Li'l Abner radar sees action in Japanese theater

These radar sets went on in a production line. We kept sending these out until August of 1945. We were sending the sets out in the summer of 1945 to the Japanese theater, probably starting in June, until August. We assembled about a hundred sets. They went to various special places.

The Li'l Abner radars were destined for the European theater. But by the time we got them finished, the European war was over. But they did see action in Japan. They were credited with shooting down some kamikaze planes and detecting some kamikaze submarines near the islands. So they were good.

After the war they were used as weather radars because they were short wavelength, X-band. You could see clouds and you could see rain on the screen. So they were useful. X-band is 9 gigahertz, 3 cm radar. That's where we operated. That was my first experience with microwaves and microwave circuits, as well as the electronics that went into the system.

Then in August 1945 came the A-bomb, and the war was over. I was still stationed at Rad Lab. In the meantime I got very interested in an invention of Pound's called the matched-T. I met a man named Richard Walker [4.24], one of the engineers, who I got to be very friendly with. We decided it would be very nice to replace the microwave RF head with a monolithic matched-T component. It's like an integrated component, microwave circuits that were built out of one solid block of metal. You'd machine it out. But we never got to build it.

CHAPTER **5**

Graduate School in Physics at MIT, 1946–1949

Joins the new Army Air Corps Cambridge Field Station

Then came the time to dissolve the Radiation Laboratory. This was September and October of 1945. It was decided by Colonel Marchetti [5.1] to form a research branch for the Signal Corps up here in Cambridge, which ultimately became the Air Force Cambridge Research Laboratories [5.2]. We were part of the Army Air Corps then, and the Li'l Abner radar program was assigned to the Army Air Corps. [*The Army Air Force would become an independent branch of the armed forces, the Air Force, in September 1947.*] Marchetti was going to be in charge of this. I guess he was a Major at the time.

He decided to recruit me to be part of this new organization. One of my first tasks was to see that as much surplus equipment as we could gather should be gathered for our new laboratory. So I was assigned to committees that parceled out equipment, and I saw to it that some of the equipment got to this new laboratory, which by the way got relocated to Albany Street in Cambridge. For the first two months, we were still at the Radiation Laboratory.

Consults for Sylvania in Boston

By December of 1945, we had the laboratory on Albany Street. I was heading up a radar group. Richard Walker [4.24] went to work for Sylvania, and I kept in touch with him. I decided that it would be a good idea for us to design an X-band RF head based on the matched-T monolithic design. So I wrote a contract, which we gave to Sylvania, where he was.

Before I left the services in February 1946, I was there for about two or three months, in the Cambridge Research Laboratory. I had put this contract in place, and I was going to go to school.

But Sylvania offered me a job as a consultant to work for them on the other end of this contract with Richard Walker. I had already been accepted to Harvard by March of 1946. I decided I could use the extra money before I went to school. So I decided to work for Sylvania for six months, on Forsythe Street in Boston, where Northeastern University is now, until the fall when I would start Harvard.

So I was consulting for them. In the middle of the summer, one of the officers, maybe a captain who was working for Marchetti, came to me and said, "Look, we have a contract with the MIT Research Laboratory of Electronics (RLE). Al Hill [5.3] is in charge of it, and we need a liaison man to interface between Albany Street and the MIT Research Laboratory of Electronics."

Accepted at MIT, becomes AFCRL liaison to MIT RLE

But Al Hill put a stipulation. He wouldn't allow anyone to be the liaison unless he was good enough to get into MIT. None of their people [at the Cambridge Research Station] could do that. They knew I was already accepted to Harvard. So they asked me if I would make an application to MIT. If I got the job I would work part time. I would get a full salary, but I could work part time as liaison man and go to school part time. It looked like such a deal! I was going to get a full salary, at that time 6,500 bucks a year, which was enormous, and I also had the GI Bill.

I made an application to MIT. In two weeks I was accepted. I had such tremendous recommendations from Millman [4.10] and from the other people. I was told some years later that Stratton [5.4] said, "If they don't give him a deal, I'll give him a deal."

It was such a good deal from the Air Corps. So then I told Harvard I decided to go to MIT. It was a tremendous deal. Here I was going to school and getting a full salary!

I spent most of my time in my classes. About once a week, I would go and tell them at the Cambridge Research Station what's going on at MIT RLE. It was a very informal arrangement. I promised the Air Corps that when I got through I'd work for them.

Begins graduate school at MIT, joins Slater's group

I started graduate school at MIT in October of 1946. I decided to join Slater's [5.5] group, which was working on the linear accelerator. I worked with one of Slater's people on the modulator.

As a student, I took three courses so I could have time to work on the project. That was my job as a liaison officer, to work on this project, which was one of the things the Air Corps supported, among other things. They also supported some magnetron work by Smullin [5.6]. He was designing magnetrons. It didn't come to fruition, but he was doing fundamental engineering research on magnetrons. I don't think he built any devices. So that was one of the projects. Two other projects I got very interested in were the spectroscopy work by Strandberg [5.7], who also was a student, but was directing the spectroscopy group at RLE, and Sandy Brown's [5.8] plasma physics group. These were some of the projects that I kept abreast of. Also, of course, I worked on John Slater's project, which was a bigger project than the others.

John Slater had a group and he used to lecture to us. We used to have weekly meetings. The linear accelerator was going to be a periodic accelerator. He wanted to solve the problem of a periodic structure. He wanted somebody who knew about Bessel functions. And that was my honors study at Cooper Union.

He asked if somebody knew Bessel functions. I raised my hand: "I know Bessel functions." This was early in the fall. He must have looked up my age and my background and decided that I was an engineer without a physics background. And here I was volunteering to do a theoretical problem.

About early November or middle-November, just a little before Christmas of 1946, I met him in the hall and I asked him, "Professor Slater, why didn't you let me work on this problem with you?" He said, "Well, you're too old to be a theorist." I was approaching 31 years of age.

I was very upset and offended. I decided to leave his group. The following spring, in 1947, I decided to join Sandy Brown's group and work with them and Will Allis [5.9] and take their courses. I took their courses and I got all As my first two years.

Gets all A's in first-year graduate courses in physics

In my first year of graduate school, I took thermodynamics, atomic physics, and theoretical mechanics for my first term. The next term I took electromagnetic theory, X-rays, and also nuclear physics with Deutch [5.10]. That was the extent of my physics courses so far. I got A's in all the courses. In fact, I was one of the few people who got an A in Deutch's course. Most everybody flunked. It was on radioactivity, and it was solving differential equations. This was so simple.

Thermodynamics was taught by Allis, theoretical mechanics was taught by Slater, and atomic physics was taught by Mueller [5.11], who was fantastic.

The first exam I took with Allis was on a wintry day, and I got delayed. I didn't have a car. I got delayed coming to the Institute that morning because the trolley cars were held up or something, and I got into the class a little late. I was so nervous by that time. I think I got 65 on the exam. It was the first time I ever flubbed an exam to that extent. I think I didn't finish because I was late. The rest of the exams I took I got high marks. So I ended up with an A anyway.

Allis taught that course in thermodynamics, and he was a good teacher. Mueller was a good teacher, and Slater was a good teacher. So my first three courses were probably with three of the best teachers at that time.

The next term I took electromagnetic theory from Slater, and I took X-rays from Warren [5.12]. I got all A's the first year.

Takes Slater's course on electromagnetic theory

I remember taking the course in electromagnetic theory from John Slater. I would go through various books and look at other problems. I discovered that John Slater himself took problems from these other authors.

He used to give problems during the week. I had funding from the GI Bill, so I not only bought his book, but I bought books by other authors. I discovered there was one particular book on electromagnetic theory that he would take problems from, vary them, but essentially they were similar problems. So I would do them. When the final came, I just went through these things, read them and did some of them. There were four major problems on the final exam, and two of the problems I had done before. So I got a very good mark, and of course I got an A in the course. It was that simple.

I took nuclear physics in the spring with Deutch and that was very interesting. Deutch wasn't the best lecturer. He gave us an exam. I did very well on it, but most of the class flunked. They went into the graduate office to protest. I didn't join them because I had no reason to complain. They weren't properly prepared for the exam.

The reason I did well is because I just don't take a course, I study. I take other books. For example, even though I took his course, I was studying nuclear physics from a book written by another MIT professor, Evans [5.13]. I used Evans' book as a guide. We didn't have a text in this course. That was part of the problem. But I used Evans' book as supplementary reading. So I was well prepared for this exam.

That first exam was tough, but I'm a good exam taker. There was a fair amount of mathematics, solving differential equations, which was right up my alley. It was radiative decay. It was part of the exam, and I suppose there were others.

Takes PhD-qualifying exams in second year

I did so well my first year that Ned Frank [5.14], one of my advisors and a partner of Slater's, was very impressed. He said, "Ben, why don't you take the PhD qualifying exams [5.15] in the fall?" And I said to him, "But I hardly had any courses in physics." He says, "Well, if you flunk, we won't count it."

I spent the whole summer on my own, since I wasn't responsible to anybody. I quit Slater's project because he rebuffed me.

That summer I kept attending Sandy Brown's group meetings, and I took two summer courses, one at Harvard on physical electronics with Ed Purcell [4.22]. I got an A in that, but it wasn't recorded because it was at Harvard. And I took a course at MIT on acoustics with Morse [5.16]. He wasn't teaching it, somebody else taught it. I took those two courses because those were the only ones available. I just wanted graduate courses.

I started studying for the qualifying exams. I studied nuclear physics, statistical mechanics, a little bit of quantum mechanics, and more atomic physics. I decided to take the exams in the fall. The exams were mostly on graduate studies, nothing in optics, which I never took, and nothing in atomic theory. I spent most of my time studying for the exams in the fall.

I spent the summer studying Evans' book on nuclear physics and other things, statistics which I didn't have, to prepare for the written exam in the fall. I was taking courses in the fall, which I neglected to some extent to study for the exam. I think I took the exam in the end of October, about a month after.

Passes written exam, flunks oral exam

Lo and behold, 30 people took the written exam, including Manny Maxwell. I don't know whether he passed it. Only ten passed it, and I was one of the ones who passed it. I had a knack for taking exams. I wasn't informed until about the beginning of December. Ned Frank came to me on a Friday and says, "Ben, you passed the exam. Your oral is next Monday, and these people are your examiners."

He didn't even give me a chance to talk to them. My three examiners were Sandy Brown, Harvey [5.17], and Hardy [5.18], an optical physicist whom I didn't even know. I don't know if it was Hardy or one of his associates. Anyway, I didn't know him from Adam or Eve. I didn't know what to study.

So the next Monday, I went in to take the orals. It was one of Hardy's younger professors who worked with him. He was an optics man. Sandy Brown was the second. The third one was Harvey, because I offered theoretical mechanics as one of my specialties. You had to offer two specialties, so I offered gaseous electronics and theoretical mechanics.

So this fellow, not Hardy but his associate, started me off and asked the first question, "What's a minus one diopter?" I said I never heard of it. He started asking me about lenses and things. If I had studied it at Cooper Union, I had forgotten all about it. I said, "Look, I haven't taken a course in optics." By that time I was very nervous.

Then they started asking me other questions. Sandy Brown started asking me questions in his field and I did all right. But I didn't do very well on the orals. They threw me for a loop. Then they asked me other undergraduate things that I hadn't taken. I would say, "Look, I haven't taken this course. I don't know." So I kept saying "I don't know" until they again started asking about my specialties.

This was the beginning of December 1947. Three weeks went by. They agonized over my case. Sandy Brown said they had a big discussion. Everybody said I was a bright guy and that I would do well. I got A's in all my courses up to then. The question was, should they pass me and insist that I take more undergraduate courses, because that's where I was failing, or should I retake the oral exam, which is what happened.

So they flunked me on the orals. I was very disappointed. I mean, here I passed the written exam.

While we were going through the hall one day in the RLE, Stratton came along with two of my buddies who had passed the oral exam and who had been there longer. I didn't start until the fall of 1946, and these people were further along than me. They had passed it, so Stratton congratulated them. Then he realized I was there, and he knew I had flunked. He apologized to me.

Retakes and passes qualifying exams

The next exam was different. Whereas the first exam I took was mostly on graduate courses, now they changed the format. It was half on undergraduate courses and half on graduate courses. It was to be taken in February 1948. This time I had about six weeks to study. I'm a very good student. Believe you me, I studied optics and some of the other subjects. I studied all the books by Sears [5.19]. I learned undergraduate physics. I had no trouble passing the written exams. But this time, I could take the orals the next fall, not right away.

You needed two specialties. One of my specialties was physical electronics, which is plasma physics, such as occurs in old-fashioned vacuum tubes; I had training in that. I studied Cobine's book [5.20], which includes both plasma physics and physical electronics. That was one of my specialties.

But I had to have another specialty. I should have taken electromagnetic theory. That spring I took an electromagnetic theory course, and a special course with Harvey [5.17] on theoretical mechanics. I decided to have theoretical mechanics as my second specialty. It was a stupid decision. I studied theoretical mechanics during the summer. Who were my examiners in the fall? They were Frank, Al Hill, and Sandy Brown.

So I prepared for the orals and took the exam. This time Sandy Brown decided to start off, so that was good. I answered his questions. Al Hill asked me about the Pauli Exclusion Principle. By that time I knew physics, so I had no trouble. Then Frank asked me a hard question in theoretical mechanics about Poisson brackets, which was mostly mathematics. I knew the answer, and I started explaining it, with the equations. I kept writing. Frank stopped me and said, "Ben, what's the physics?" I said, "This is the physics." I didn't appreciate the physical meaning of Poisson brackets. I said that it's a mathematical concept. So there is a physical interpretation to it, which I guess I didn't know.

It wasn't until years later, as I got to be more experienced, that I realized there is a difference between physics and mathematics. I considered physics as applied mathematics. But then, as I got more into it, I realized the conceptual ideas and the intuitive ideas, the physical ideas could come first, and then you do the mathematics. That's what we're doing here, and that's what I did later at Lincoln Laboratory a great deal. But I didn't appreciate it then.

I passed the orals.

Discovers cyclotron resonance in the theory for gaseous breakdown

In the meantime, during that summer of 1948, before I took the oral exam, I decided what my thesis project was going to be. I had proposed one earlier to Allis, which he refused. He said he didn't want me to do a theoretical thesis. He said I was too old.

I decided to do breakdown of gases in a magnetic field. This was inspired by lectures during the spring of 1948. I listened to an Allis lecture on developing the Boltzmann transport theory for magnetic fields. I decided to do a classical theory, the equivalent of what he was doing, and I discovered, in doing the theory, that there was a cyclotron resonance phenomenon in the breakdown.

I showed this to Sandy Brown in June of 1948. He said he didn't believe it. It wasn't the Boltzmann theory. Later on, I showed you could derive that by the Boltzmann theory by making a simple assumption that was well justified.

Begins thesis experiments, confirms his theory

I assembled equipment that summer and decided to do breakdown of air in a magnetic field. The experimental curves looked almost exactly like the theoretical curves I drew earlier for Sandy Brown.

So, in August of 1948 I demonstrated the first microwave cyclotron resonance experiment in the world. At the end of August, Brown was convinced. He said, "Ben, you were right." That's what made me decide to do that for my thesis.

While I was studying for the orals, I designed a better system, and I also had cavities made at the Air Force Cambridge Research Station. I started studying the Boltzmann transport theory that I was going to do with the magnetic field, which would have been a major project, much more difficult than the experiment. That's where it stood in the fall after I passed the oral exam. By December of 1948, I was ready to do my thesis experiments.

After I passed the orals in December of 1948, I had all my new apparatus constructed, and I was ready to run my experiments, because I had already demonstrated the phenomenon in air. The experimental setup would consist of surplus equipment. I got a magnetron magnet that was used at the Radiation Laboratory as my table-top magnet. The microwave cavity I had designed myself. The microwave plumbing was surplus plumbing, also from the Radiation Laboratory. The only new thing that I needed the Research Laboratory of Electronics to do was the glass blowing of my system. They blew a glass system. The vacuum system had a fore pump, an ion pump, and an ion gauge.

I had done an apparatus during the summer, but this was now a much more refined and better system, as well as being a microwave system. The system in the summer I just lashed together arbitrarily.

During the summer I used air, so I didn't bother to even bake the cavities. To do a really fine experiment, which is what Sandy Brown insisted on, you have to bake the cavity for about a week. You get all the gases from the walls. We had a cold finger, so it would absorb all the gases. The cold finger was a buffer between the fore pump and the ion pump.

So you went to a very low pressure and then you back-filled in. Instead of doing air, this time I was going to do helium, which meant I had to outgas all the oxygen and any other impurities. And you had to control the pressure.

We had a pressure gauge, an ion gauge, that told me what the pressure was, and how much helium. We also had a little mercury inside the system. When the magnetron ionizes the helium, the helium forms a metastable ion. We didn't want that. But if you put mercury in there, it quenches the metastable ion and you have just the normal ionized helium atoms. That's what we were looking for. The ionization energy was pretty high, of the order of 20 eV. That's pretty high.

The phenomenon that I really demonstrated in my phenomenological theory, which I had showed to Sandy Brown, involved a tensor medium. In a magnetic field, a plasma tensor has an off-diagonal non-reciprocal term. But the diffusion to the walls is also a tensor, so the process of breakdown involved both the resonance and the diffusion phenomena. It was complicated.

I didn't make a quantitative check with my summer experiment in air. I just demonstrated the resonance phenomenon, which is what the purpose was. But now, with the new experiments, the question was to how make a highly quantitative comparison.

This time, I made many more cavities, of different dimensions. They were flat plate cavities, so you could treat them as two flat plates and calculate the effect of the magnetic field on the diffusion and the breakdown. That was the purpose. The title of my thesis [P5.1] was "The effect of magnetic field on the breakdown of gases at high frequencies," but the thesis really was about the cyclotron resonance phenomenon, plus demonstration of the magnetic diffusion.

Data confirm prediction of cyclotron resonance in breakdown

So we did it with helium. The difference between helium and air, of course, is that air is a molecule while helium was an atom. In helium, it turns out that the scattering of the electron by the atom is practically independent of the energy. It turned out, as later I showed both from the Boltzmann theory and the phenomenological theory, that this approximation made the two theories equivalent. Therefore, my simple phenomenological theory and the Boltzmann theory both explained this thing quantitatively. But Sandy Brown insisted I do the Boltzmann theory.

It took me just a couple of months to do all the experiments. In addition, I did another experiment. The magnetic field was initially perpendicular to the electric field in the cavity (the Voigt configuration). I also did the Faraday configuration, where it was parallel. That demonstrated the magnetic diffusion alone. That didn't show the resonance. So only the perpendicular component was resonant.

There I was, about March 1949. I had all the experiments done. But I was faced with the possibility of doing a very complicated Boltzmann transport calculation of the phenomenon.

Now, my predecessors in Brown's group included Alec MacDonald [5.21, 5.22, 5.23], who measured the high frequency breakdown in helium and hydrogen, but without a magnetic field. As you measure the breakdown voltage as a function of pressure, you get what is called the Paschen curve [5.24]. Here, it was at microwave frequencies, but it's similar. That's what Alec MacDonald did. He also did the full calculation with the Boltzmann transport equation, without a magnetic field.

But Sandy Brown had an idea. He used MacDonald's experimental data at high frequencies, and compared it to the dc experimental data.

If you take the dc data and convert it to the ac case with an equivalent dc field, which was Brown's concept, you could correlate the two, the ac and dc cases, and they would agree. So he asked me, can I do this with the magnetic field and compare it to the ac or the dc breakdown, say the ac breakdown like McDonald. I said I didn't know.

I went home and in just one evening I figured it out. I didn't have to do Boltzmann theory because Alec MacDonald had done all the Boltzmann theory, for zero magnetic field. All I had to do was identify the equivalent point on his graphs and then convert back and figure out what the breakdown was. In two weeks, I had all the theory and experiment done. I didn't have to do the Boltzmann calculation.

Completes thesis in one year

By the beginning of June 1949, I was all done with the experiments. I had plotted the data, drawn up all the diagrams, and everything. I was done. It took me, experiment and theory, just about six months. At the end of June, I was ready to write the thesis, which I did in July and beginning of August. I did all my illustrations. I was a draftsman, although some of them were done by RLE. I had Blossom, who was a good technical typist, type up my thesis. We had a typewriter. By the middle of August 1949, I was done.

So the thesis took me exactly a year. By the fall I got my degree. I started in October of 1946. I got my degree in August of 1949. It was less than three years.

I didn't take a lot of courses, but I did study a number of things on my own during that time, like nuclear physics. Even though I only took an introductory course in nuclear physics from Deutch, I learned a great deal from Evans' book [5.13].

I only took the introductory one-term course in quantum mechanics, which was a mistake. I should have taken the second term. I started taking the second term, but I dropped it because I was too busy doing my thesis.

At the end of August 1949, I did my thesis defense. I finished my PhD in less than three years, which was phenomenal. However, I don't think it was the best thing. I should have taken many more courses in physics, which I later studied on my own. But I was 33 years old, and I thought, gee, it was time to start doing a real job, although I did have the part-time liaison job with the Air Force.

Inspiration for thesis came from an Allis' lecture

Allis said his most significant contribution was educating students. I must say, I learned a lot more from him than I learned from Sandy Brown.

Sandy Brown was in charge of the laboratory, but I never saw him in the laboratory. I learned how to do things from my predecessors, like Mel Herlin [4.21] and others, who already had set it up. I decided to do the magnetic experiment on my own. Sandy Brown didn't think it was going to work.

Of course, in two months it worked. Then he agreed it would be a good thesis topic. I went on from there myself. And I never consulted either Sandy or Will Allis. From there on I did the theory myself.

I got the inspiration from one of Will Allis' lectures. Allis was doing the Boltzmann transport theory in a magnetic field, and I became very curious about what it would do. Instead of doing the Boltzmann transport theory, I did the phenomenological theory. That's what Sandy Brown doubted. He felt the Boltzmann theory was the only way to calculate it.

But later on, when I did my thesis, I showed the two were equivalent under the conditions we were doing experiments, because we were doing it in helium, and, at the energies we breakdown helium gas, the scattering time was independent of energy. If you do that, you can take the conductivity integral outside the integral, and then it becomes the classical expression. But if the scattering time is a function of energy, you can't do that. Then it's more complicated. I was thinking, gee, that's going to be too difficult.

Fortunately I didn't have to do that. I showed the two were equivalent, that the phenomenological theory explained everything.

It was a very interesting lecture that Allis gave, which inspired me to put a magnetic field on the breakdown. And that was my thesis.

Allis was a lovely person. He had a nice life. We students all were very fond of him.

I would say the most interesting thing that Allis taught me was his enthusiasm for physics. It was there even in the thermodynamics lab that he taught me, and then of course in my interaction with him on gaseous electronics. We all attended all the group meetings, so I had quite a bit of interaction with him. He was one of the pioneers in doing Boltzmann transport theory for gaseous electronics. The other one was Margenau [5.25]. But I think Allis' theory was more complete and more significant.

The new thing that Allis was doing, that hadn't been done before, was to adapt the Boltzmann transport equation to the microwave regime, the time varying case.

So you solved the Boltzmann transport theory in the time-dependent domain, and you got the expressions for the conductivity as a function of frequency. That's what makes it more complicated. Allis also showed that the breakdown equation led to a differential equation, whose solution was a geometric function, one of the more complicated analytical functions. I think the guy who did most of the work on it was Alex MacDonald, a Canadian and one of my predecessors. He was a year ahead of me.

And I said to myself, gee, solving that time-dependent case in a magnetic field would be a horrendous thing to do. But when I showed that the Boltzmann transport theory was equivalent to the classical, it became a much simpler matter. I didn't have to do that. All I had to do was transform Alex MacDonald's data from the non-magnetic case to the magnetic case, using the concept that Sandy Brown introduced, the equivalent electric field. But he didn't do it for a magnetic field. He compared the microwave case to the dc case, and you can show that the equivalent electric field involves the frequency.

In my case, it involved something more complicated. But I was able to make that transformation. In a week I interpreted all the experiments by using the theory and data of Alex MacDonald, and transforming it to the equivalent magnetic field case. It agreed with the experiments beautifully, just as Alex MacDonald did with the zero magnetic field case.

You see, what you do in plasma physics is look at the Paschen curve. That's the breakdown field as a function of pressure. That's what Alex did, he varied the pressure. He did the microwave case. You could also vary frequency, but we did it at one frequency. We all worked at S-band, which is 10 centimeters. But now, if you vary the frequency, then you also explore the Paschen curve. It also has a minimum. It's all I had to do, use the equivalent in the magnetic case. It just was a transformation of variables and it worked beautifully.

I would say that was Sandy Brown's major contribution, that he conceived of the idea that you could transform from the dc case to the microwave frequency case. Then he asked me, can I do the same thing to the magnetic case, and I said I think I can. In one evening I worked out the theory.

I think Will Allis, his course and discussion, and the study sessions and the group sessions with him there, was, I think, the big influence. I would say he was the intellectual leader of that small group that was doing microwave studies.

I suggested a theoretical thesis but he didn't want to take me on. He was the one who said, "You're too old to do theory." I ended up doing both. I picked something that turned out to be much better. It got me into the thing that I built my career on. I think that was a fortuitous and happy choice.

But I give Allis the credit for inspiring me through one of his lectures. I would say that his course made a big impact on me. I have since used the Boltzmann transport theory for a lot of things. In fact, when Zeiger and I interpreted the semiconductor cyclotron resonance, we used the Boltzmann transport theory. I was the one who suggested that as one possible approach, although there are other approaches that work too.

Writes paper on thesis

I wrote up a paper on my thesis. I wrote it out, very straightforward, very simple, like I do. Sandy Brown and Allis modified it. I wanted to send it to *Physical Review*. They didn't want to let me send it to *Physical Review*. I should have sent it on my own. I don't know why they did that. The work was highly original. Sandy Brown used to present it as if it was his own work, which it wasn't. He and Allis had very little to do with it.

I can say this honestly now. I picked the thesis. I did it all by myself, with a little consultation now and then with Sandy and the others, with Sandy on the experiments. But essentially I learned the experiments from the other students, and for the theory, I didn't even use the Boltzmann theory. I did it all by myself. But they wanted to modify it, and the modification was trivial.

I think what I did was much simpler and more straightforward. In fact, the way I wrote it up was the way I wrote up cyclotron resonance in germanium later on, which is the way to do it. Simple is the best.

They insisted on publishing it in *Journal of Applied Physics* [P5.2], which was a mistake, because this was the first microwave cyclotron resonance experiment. Had it been published in *Physical Review*, it would have been much better known. But at that time I was naive and didn't know. I accepted it.

Graduate courses in physics at MIT

The first semester of quantum mechanics was just the Schrodinger equation, and you solved some simple problems. One of the problems that Feld gave on the exam that I didn't know how to do was the Zeeman effect, which I must have missed because I skipped some classes because of studying for the exam. That was the only B I got while I was there.

But the second term, where I started taking the course and I dropped it, and I didn't even sit in on it, they gave you more advanced things: angular momentum, orbital momentum, spin. If I had taken longer to get my PhD and taken two years of courses before I took the exam, which is the normal thing that people do at MIT, I think that would have been a much better thing.

Many of the things that I missed I had to learn subsequently. For example I didn't take the physical electronics course that was given by one of the professors at MIT. But I took the equivalent course, a much better one, from Purcell [4.22] at Harvard.

There was no such thing as a solid state course at MIT. I didn't really learn solid state until I came to Lincoln and sat in on Harvey Brooks' course at Harvard. Slater taught a theoretical course on band structure, but that wasn't the kind of solid state you needed. Harvey Brooks' course was more to the point. It provided us with a good background and that's where I learned quite a bit.

Although I only went three years, subsequently I learned a great deal of plasma physics working for the Air Force. I learned about ionospheric physics, magneto-optical theory, you know, that Appleton and Hartree did. I learned magneto-hydrodynamics on my own, although I never used it.

I learned more about the Boltzmann transport theory from an English book, which included magneto-hydrodynamics, and which was more advanced than the course that Allis gave. So I learned a great deal there, and I learned more quantum mechanics on my own, about the Dirac equation and so on, and a lot more about perturbation theory, which I needed to interpret my experiments.

CHAPTER 6

Postdoctoral Work at Air Force Cambridge Research Laboratories, 1949–1951

Joins AFCRL

In September of 1949, I decided to join the Air Force Cambridge Research Laboratories (AFCRL) full time. There was a group there that was doing plasma physics in the Geophysical Laboratory. A fellow named Oz Fundingsland [6.1] wanted me to join his group. So I decided to do that, although I could have just started my own program on Albany Street. They were located at the Watertown Arsenal in Watertown, Massachusetts. I decided to join his group.

Captain Marchetti [5.1] was the Director of AFCRL. Marchetti was quite an operator. He wasn't a scientist; he may have had a degree in engineering. He was more of a radar engineer than a scientist. But he was a very intelligent man, a very energetic and shrewd man. He recruited people like me and others to work there. He had some good people like David Atlas [6.2], who started using X-band radar to do weather.

They had a decent program. But many of the people were not top notch, unfortunately, as I found out when I got there. Even some of the ones who were good were more interested in operating and trying to get promoted in Civil Service than doing their jobs.

At any rate, I joined the Geophysical Laboratory at AFCRL. In the space science area, there was a theorist who was working on ionospheric physics. I've forgotten his name, but that was very interesting. And there was this plasma physics program.

Measures tensor susceptibility of a plasma

I started my microwave experiments. I built my own setup, and I did some more sophisticated stuff on cyclotron resonance phenomena at microwaves. In order to interpret the experimental results, I had to use the perturbation theory that was developed by Hans Bethe and John Slater [5.5] to figure out how the plasma inside the cavity perturbed the resonance frequency and the Q of the cavity. This is what enabled me to calculate the breakdown fields.

Then, when I learned magneto-ionic theory and more about tensor media, I realized that I could generalize the perturbation theory to tensor media. I think I was the first one to ever write the perturbation theory for a cavity with tensor media. My new experiments depended on that.

This time, however, I did the experiment differently. Instead of studying the breakdown, I studied the decay of the breakdown. What I did was break the cavity down in a magnetic field and let the ionization decay. So I had a two-frequency system, one frequency to do the breakdown, using a magnetron and a klystron, and a weaker frequency to probe the resonance as a function of density and time.

Essentially, I was also measuring the decay time at the same time. There were long decay times because of the metastable helium, so it gave me plenty of time. I figured out, before I did the experiment, that if I were to put this tensor medium in a magnetic field, I would split the resonance of this cavity.

I particularly picked a cavity with a TE-111 mode, which is a circular mode. It's a linear mode, but it's really composed of two oppositely circularly polarized modes. One would shift one way, one would shift the other way, and therefore you'd see two peaks. And indeed I did. It was such a wonderful thing to have predicted this thing theoretically and then see it!

That essentially showed me that I'm measuring the tensor susceptibility of the plasma. This, of course, later on, was what gave me an idea of the experiments that we did at Lincoln Laboratory some years later on ferrites.

I had developed this theory, and in fact I think we presented some data at RLE.

I wrote a paper on all this and I sent it in for publication. The referee refused it! He said this sort of phenomenon occurred in the ionosphere. He was all wet. I don't know who the referee was, but apparently he was prejudiced and ignorant, because what I did was quite fundamental. I didn't bother to resubmit it. Thirty years later somebody published a paper in *Physical Review Letters* with a result of far less quality, not even referring to my work.

Calculates ambipolar diffusion in a magnetic field

In the meantime, I started working with Oz Fundingsland [6.1], both on his project and my project. I was learning a lot of ionospheric physics.

I was putting several people to work because I had so many ideas. I put people to work doing computations. I worked with George Austin [6.3], who was a captain in uniform. He was a good physicist. I think he had a master's degree. He and I were the first to derive the formula for the ambipolar diffusion in a magnetic field. We didn't publish it. Some years later, Simon [6.4], a well-known plasma physicist, published it, and he got credit for it. But we had it. We were just having fun, and I guess maybe we didn't understand the importance of publications.

Ideas on toroidal plasma

Oz Fundingsland built a toroid and was trying to do ion cyclotron resonance. It didn't work. But it taught me a great deal. Later on, I figured out why it didn't work. That was the beginning of the idea of a toroidal plasma, that you could confine it more. So it wasn't original necessarily with the Russians. We were doing it.

We used the dc breakdown and tried to use the coils to detect ion cyclotron resonance. The problem with that was we didn't go to low enough pressure and the plasma ionization was too high, so we couldn't penetrate the plasma. But we didn't realize it then. I didn't really learn plasma physics until I began to understand ionospheric physics, but that's another story I'll get to.

I worked at AFCRL for two years. The first year I was doing my own experiments, and then I was working with Austin and with Fundingsland.

Fortunately, the rocket exploded

One of the division leaders had a project. He was interested in electron cyclotron resonance in the ionosphere. There was a well-known physicist, who had wanted to break down the ionosphere from the ground by putting a whole bunch of antennas on the ground at the cyclotron frequency of the earth's magnetic field. This is roughly at about a megahertz, so you have big antenna arrays. He would send this up to the ionosphere and, very similar to my experiment, the breakdown field is reduced considerably at those very low pressures. Well, nobody ever built that. That was an expensive proposition.

This division leader got the idea that he would do it by sending up a rocket with an antenna at the tip of the rocket, and he would light up the sky. It would be a Roman candle. He asked me to study the project. I told him the thing was ridiculous. I told him that, if you have an antenna in which the electric field is maximum, since it goes inversely as the radius, all the breakdown would occur at the antenna and would break down to the body of the Aerobee rocket [6.5]. He didn't believe me. I said it wouldn't light it up. It would be just a spark, and it would shield itself.

He didn't believe me, so I told him, "We'll build a little apparatus." We did build a vacuum apparatus, and what I thought would happen did happen. He didn't want to hear about it. He had already spent a quarter of a million dollars at Tufts University to build equipment on an Aerobee rocket. He was going to fly it. That's the kind of scientist he was.

This didn't help my political position. He took away some of my help. I told him that experiment was nonsense. Here I am, a young scientist. He appreciated me. He told me, "You're a brilliant scientist." He saw what I had done in the lab and he was pleased with it. But he didn't like the idea of my throwing cold water on that experiment.

Anyway, they went ahead. It was very fortunate that the Aerobee rocket blew up. This is the kind of people who ran experiments. This is one of the things that disillusioned me.

Becomes interested in fusion

About that time, I met two people who were consultants for the Air Force, Lou Gold and Fisher. Fisher [6.5a] was a man who was very interested in fusion. So I first got into this area through these two, I think in early 1950. I was very excited about it.

I started studying fusion. They were going to do it with sparks and pinches. I got the idea that it was best done with magnetic confinement, so I started studying magneohydrodynamics. I started studying very sophisticated Boltzmann transport theory.

I also started listening to Menzel [6.6]. Menzel was an astrophysicist at Harvard. He was very much interested in solar physics. AFCRL was supporting his program. I would go listen to his lectures, and that's how I got introduced to magneohydrodynamics. I was very fascinated. I learned the Boltzmann transport theory. There was a very sophisticated book on it that I studied. This taught me a lot of Boltzmann transport theory. In the meantime, from ionospheric physics, I began to learn about plasmas. I started talking about plasma waves. Again, I used the Boltzmann transport theory and showed the equivalence between that and the phenomenological theory, and I calculated the dispersion equations for a simple plasma.

It was about the time that Bohm [6.7] and Gross [6.8] (Gross is a friend of mine, he's at Brandeis) solved the Vlasov equation [6.9] and got the dispersion relation for a hot plasma [6.10]. I was using it for a cold plasma. I gave lectures. I gave a lecture to Sandy Brown and Allis [5.9]. I started talking about it. They didn't believe in plasma waves. But Eugene Gross attended that lecture. I was calculating the tensor components and all the dispersion relations of the cold plasma, with the Faraday configuration and the Voigt configuration, on which Allis, who didn't believe all this, later wrote very authoritative papers. Which is amazing.

There was an intellectual difference between me and Sandy Brown. Sandy thought I was very cocky. I didn't think he was very bright. I thought I was brighter, which turned out to be the case. He couldn't buy the idea that a fellow like myself, who didn't have a physics background and who didn't pass the orals the first time, could be so creative and original. He always thought I was an overachiever. But he was wrong. So there was an intellectual gap between us.

Allis' problem was that he looked at things in a very complicated and mathematical fashion. Sandy Brown didn't understand theory, and I was somewhere in between, where I could do the simple theory and understand the physics and understand Allis too. But I thought very often that his was more mathematics and a little less physics, although Allis was a good physicist. I won't deny it. But I had a better feeling for the overall picture than either of them.

After I did my postdoctoral work at AFCRL, I got into fusion and other things. I got interested in magneto-hydrodynamics as well as fusion. I became a pretty good physicist at that, and I was working in these fields.

When I gave my lecture, Eugene Gross says, "You neglected one term, the Vlasov term with the magnetic field," which is the term that gave you the dispersion for the hot plasma. He decided to go home and solve the problem, which disappointed me. I wanted to work with him, but he didn't want to. He solved it. He was the first one to solve the integral-differential equations. He did it wrong. Later, it was done by Ira Bernstein [6.11] and they became the Bernstein modes.

I knew what he was doing, but I didn't know how to solve the equations. He knew. It was too bad, because if I had solved the equations, this would have been a real coup. But at least I gave them the inspiration to do this.

So, here I was, a believer in plasma waves, but I didn't fully appreciate the distinction between a hot and a cold plasma until I read the Bohm and Gross paper [6.10].

Decides to leave AFCRL, rejects Los Alamos job offer

By that time I had decided I had had enough of AFCRL. I didn't like the environment. The people weren't real scientists. They didn't work that hard. It was more important to jockey for favoritism with the bigwigs, with the director, rather than do the work. So I decided I was going to leave.

I looked for two jobs. I made an application to the MIT Lincoln Laboratory, and I also flew out to Los Alamos. I told the folks at Los Alamos about my ideas on fusion, magnetic confinement. They offered me a job as a plasma theorist. After I was there, I decided this wasn't the place for me, particularly because Blossom, who was a city girl, wouldn't care for it. She'd be isolated. So, I decided to take the job at the MIT Lincoln Laboratory. That's what made me decide. That fall of 1951, before I joined Lincoln Laboratory, I attended a conference on plasmas and gaseous discharges. I gave a paper on my theory of tensor media and plasmas in a cavity [P6.1, P6.2].

Another thing disillusioned me. I had already made up my mind that the way to do plasmas, fusion, was in a magnetic field. I think I talked to Allis, but more important, I talked to Ira Bernstein [6.11], who hadn't yet made his name, and to Ted Holstein [6.12], who were two of the promising theorists. Ted Holstein already was established. I told them that the future of plasmas was with magnetic fields. They laughed at me. I decided, well, these are the people who are considered the leaders of the field, and there was a lot of political jockeying among these people. They were the big shots, operators. I was a brash young man. I considered them inferior. I said (to myself), these guys really have no vision or imagination.

But the idea was to use magnetic confinement and magnetic fields, and I was studying magneto-hydrodynamics at the time. I think it was the right approach. Ultimately, those were the things that ultimately evolved and were developed in the fusion program.

But there I was in 1950–1951, predicting where the future would go in plasma physics. The only people who would listen to me were at Los Alamos, and I didn't want the job they offered me. I wasn't a plasma theorist. Allis said, "You can't be a theorist." But these guys were convinced I could be a theorist. I had taught myself theory during those two years at AFCRL. I'm sure I would have done well at Los Alamos. But I didn't want to go there.

The other thing that scared me is when I flew, for the first time, from Albuquerque to Los Alamos. From Albuquerque we flew in a single-engine plane. There I was, looking down, and there was this mesa. Just this runway and then a drop. I said, "Gee, if you miss that we're in trouble." I didn't like the idea. In the future, when I went to Los Alamos, I never went by plane. I drove up from Albuquerque.

So I decided to quit AFCRL, go to Lincoln Laboratory, do something else, and work at MIT where you had good scientists. It was a wise decision.

CHAPTER 7

MIT Lincoln Laboratory, 1951–1965

7.1 Beginning of a Scientific Career

Negotiates offer from Lincoln Laboratory

I got an offer to go to work at the MIT Lincoln Laboratory [7.1]. I finally ended up in Jerry Wiesner's office [7.2]. I had already decided to join Lincoln. We were negotiating the terms of the offer. He was at that time, I think, head of the MIT RLE, or at least one of the assistant heads, I guess, assistant head to Al Hill [5.3]. But also a big shot in the planning of Lincoln Laboratory. He was on some sort of committee.

He was the one who decided to make the offer. I told him what I was making at the Air Force and he offered me the same salary. I said, "Look, I'm worth more. I now have had more experience." He said, "Take it or leave it." I had already made up my mind to go to Lincoln anyway, so I took it. I think the salary was either $6,500 or $7,500 a year. But other people, I think, were getting more with fresh PhDs, with much less experience than I.

But I took the job. At that time there was to be a Solid State Group, and I met you [*Don Stevenson*] and Dick Adler [7.3], and I decided to work with you. But I didn't have my clearance yet, so instead of being in Building 22, I was in Building 20 for a while.

Mentors MIT EE doctoral student

I was in Building 20 for a few months, and there I met an MIT graduate student, Berk [T53], and he was looking for a thesis topic. He was in the MIT Electrical Engineering Department. I told him about the work I had done at Air Force Cambridge Research Laboratories (AFCRL), and that I wanted to extend it. It would be very good work. I directed his thesis. He got his thesis [T53], and we presented some work at an IEEE meeting involving these phenomena [P7.1, P7.2].

Lester Hogan visits MIT, lectures on ferrites

It was about that time that Lester Hogan [7.4] visited MIT to talk about ferrites. I immediately realized that [*the mathematical properties of*] ferrites were very similar to the plasma tensors except, instead of electric susceptibility, it was magnetic susceptibility. So, I got the idea of doing both the perturbation theory, for not just the dielectric susceptibility, but also for the magnetic susceptibility. I extended the perturbation work I did at AFCRL for ferrites in a cavity.

Idea for ferrite experiment with Artman and Tannenwald

Then I got the idea that that would be a wonderful way of measuring the tensor, and I proposed this experiment to Artman and Tannenwald [7.5], and we started the experiments. So I planned that experiment. That was one of the first things that I planned on ferrites. It was still within the Solid State Group.

In the meantime, I was studying a great deal about semiconductors. I was reading Shockley's recently published book [7.6] to try to understand semiconductor junction devices, because the group's primary mission, according to Zacharias [7.7] and company, was to learn as much about the transistor as possible. Zacharias was the spark plug for starting semiconductor research at Lincoln. Of course, we also had to learn about the fundamental properties of semiconductors.

That's how we got started. I was working on two sets of programs, mostly theoretical, and occasionally I gave a lecture on the Hall effect and other things. I understood semiconductor theory pretty well and picked it up quickly, but I also began developing ideas on ferrites. Then, when Ken Button [7.8] arrived, I got the idea for solving the electromagnetic problem of a ferrite slab in a rectangular waveguide. That, along with the Artman–Tannenwald work [7.5], was the beginning of my ferrite work.

Decides to take charge

At the same time, while I was at the Air Force Cambridge Research Laboratories, I made the decision to stick just with my science, but I got into trouble by not playing the politics and not taking primary responsibility. I decided the next job I have I'm going to take charge as much as I could.

At Lincoln Laboratory I saw a big vacuum, and I don't mean to insult you [*Don Stevenson*] or Adler. But I thought that I could – let me put it in a positive way – I thought I could help you. The group was growing so big, I thought I could help you and Adler in interviewing people, in helping to set up programs. So although I was officially not an assistant group leader, I actually worked with you much more closely than with Adler. I helped you in interviewing people, helped to set up programs, and coordinated people. In fact, people would come to me and ask me what they should be doing. And I would suggest ideas to them. But you and I decided that we were going to work a little more closely together.

Visits Lark-Horowitz and Fan at Purdue, gets idea for cyclotron resonance

I visited Lark-Horovitz [7.9] at Purdue University, and we also saw Fan [7.10] there. Fan was doing microwave experiments on semiconductors. That's the day that I got the idea that I wanted to do the same experiment, but in a magnetic field. This was the beginning of the idea of doing cyclotron resonance.

Unfortunately, since already I had started the ferrite work, I couldn't convince Adler to buy another magnet to do this [*cyclotron resonance*] work. And, I think that summer or fall, Adler decided to sacrifice me for some reason to the "crash program," instead of insisting that I stay with our program, the solid state program. Perhaps Al Hill asked for me. Anyway, Adler agreed that I would work on the "crash program."

Transient response of a p-n junction

There was another problem that I got the idea for. It may have been during the summer after I joined Lincoln. When I went to Los Alamos, I got the idea, from something they were doing, of solving the problem of the transient response of a transistor, or of a p-n junction. So I started a program with one of the mathematicians at Lincoln, Siegfried Neustadter [7.11], to calculate the transient response of a p-n junction [P7.3].

Works on four projects

So, I was working on four ideas: two on the ferrites, one on the transient response of a p-n junction, and I was starting to think about doing cyclotron resonance. In fact, I think that was the time I hired Zeiger [7.12], and I was talking to him about the idea of what I wanted to do. I think we called it diamagnetic resonance.

But then all this was interrupted by the "crash program" in the summer of 1952 that I was asked to work on. That took up most of my time. Ken Button [7.8] was also asked to work on the "crash program" because we both had radar experience. We were probably the only two in the group with radar experience, and they needed people.

Becomes leader of Ferrites Group in 1953

The "crash program" was successful, and my bargain with Bill Radford [4.6] was that if it's successful I get my group. In the meantime, I had met Bob Fox [7.14], and since I was working already in ferrites, I thought it was a good idea to start an applied effort in ferrites and build the group around that. Essentially a microwave group, because I thought this was needed.

So we proposed this to Radford [4.6] and he approved it. In 1953, May or June, we became a group, and I transferred Zeiger [7.12], Tannenwald [7.5], Foner [7.13], Button [7.8], and myself. Those were the key people. Bob Fox [7.14], Jim Meyer [7.15], and Dexter [7.16] also joined the group. So that constituted the group, and we were primarily a Ferrites Group.

7.2 Cyclotron Resonance

Begins cyclotron resonance experiments

Then in May of that year, 1953, Shockley published his theoretical proposal on the cyclotron resonance in a paper in *Physical Review* [7.17]. Since I had already thought about how to do it and figured out the technique for doing it, I decided that his scheme wouldn't work. That summer of 1953, I asked Zeiger to collect equipment, because we were going to do this!

Invites Kittel to lecture at Lincoln Laboratory, rivalry begins

During that summer of 1953, I invited Kittel [7.18] to give us talks on magnetism, because we were working in that field. He gave us talks here at

Lincoln, and at that time I told him we were planning to do cyclotron resonance. He looked me and Zeiger in the face and said, "Well since our friends at Bell and here are doing it, I guess we're not going to enter it." He was lying [7.19, 7.19a].

August came around. Two months went by. I asked Zeiger, "Did you order the equipment?" I was busy organizing the group. He said, "No." I said, "Why not?" He said, "I don't think your idea will work."

I got very upset. I decided to recruit Dexter and work with him. We planned the experiment. By September 1953, we had the equipment. The experiment depended on the breakdown. [*Ionization of the impurity levels by breakdown was necessary to produce free carriers that could participate in the resonant absorption process.*] The key to it was, Shockley [7.17] said he would do it at 20 degrees Kelvin, and I knew he wouldn't freeze out the carriers and that he would run into the plasma effect, which would prevent him from doing the cyclotron resonance properly.

Cyclotron resonance experiments are successful

By that time I understood that, if the plasma is too dense, you can't penetrate it. So my idea was to freeze out the carriers and re-excite them with the microwave and do the analogous experiment that I did for my PhD thesis, except on a semiconductor, and look for the resonance breakdown. That was the fundamental idea. And of course, when we did it, it worked.

However, when we did the experiment my microwave power was too high, the sample was too large, so that essentially the high power in the microwaves broadened the line and we didn't see it.

But I already planned the anisotropy experiment in which you rotate the sample relative to the magnetic field direction. When I rotated the sample, even though we were studying the breakdown, we could see that the bumps [*in the plots of absorption versus magnetic field*] changed, in other words, shifted in frequency, but were not resolved. So we knew there was anisotropy. This was on n-type germanium.

Then, through the grapevine we heard that Kittel had done cyclotron resonance. I went to Harvey Brooks [7.20] and told him, "I'm going to proceed. I don't think they did it right." I told that to Earl Thomas [7.21] who was the group leader after Adler. They both said, "How can you compete with Kittel and company?" I just said, under my breath, "To hell with you. I'm just as good as these people." I decided to go ahead [7.22].

Improved setup detects anisotropy of electron mass in germanium

So we revamped and rebuilt our setup to make it more sensitive and use less microwave power. And, we were the first ones to publish data for the anisotropy of the effective mass of electrons in a semiconductor [P7.4].

Then, through a suggestion from Burstein [7.23], I got the idea that we could use light excitation, which could require less microwave power and make it more sensitive. And that's how we were able to observe the anisotropy of the valence band [P7.5].

So we were the first ones to measure the anisotropy of the cyclotron resonance masses of germanium, both the holes and electrons, and determine their quantitative properties. We published these as a series of Letters to the Editor in *Physical Review* [P7.4–P7.8].

Kittel credits Lincoln Laboratory's results in his invited paper

That January 1954, Kittel was scheduled to give an invited paper [7.24]. I was very upset with him for lying to us. So I made sure, when we submitted our paper on cyclotron resonance, that I wrote to the editors that it was not to be refereed by Kittel, but by somebody else. I guess they did that. Anyway, I met Kittel the night before he was to give his talk. By that time our preprint [P7.4] was accepted for the holes and electrons. I put it in his hand, and when he saw it he went white.

The next day, when he gave his talk, he announced our results, and overnight I became famous. I was the first author on that paper [P7.4], properly so because it had been my idea. So we were the first ones to do really a complete experiment on cyclotron resonance in a semiconductor. And of course once we got the light excitation ... I'll talk more about that later.

Why early cyclotron resonance experiments at Lincoln Laboratory were successful

Well, there were several elements that contributed to the success of the cyclotron resonance. First, of course, was the key idea – which I had – to use microwave excitation, the freezing out of the carriers and using microwave excitation. And building a paramagnetic resonance spectrometer, which Dexter assembled under my supervision and direction. This was a standard type of spectroscopic apparatus that a number of people around the country had. We just merely copied the design in the literature.

The other factor that was important was to get hold of good germanium. Now I had decided that the key to the experiment was to use doped germanium, both p-type and n-type, not heavily doped but doped enough so that the impurity scattering would be minimal. The idea was to freeze out both p-type and n-type, and then re-excite it with the microwave, ionize it in a controlled fashion, and then sweep the magnetic field and see the absorption as a function of magnetic field.

The key was to get good germanium with the right amount of doping. Homer Priest [7.25], who had spent the previous two years in growing germanium, was able to supply me. These were very important contributions by him.

So when we did do the experiment, it worked like a charm, except as I said before, we hit it with too much microwave power, and so the lines were broadened. We designed it so that we had a wire going down to the germanium crystal, and we could rotate it relative to the magnetic field and look for the anisotropy that Shockley [7.17] predicted might occur.

Indeed, for the electrons we saw – it was n-type germanium we did first – we saw the bumps [*in the absorption versus magnetic field data*] but we didn't resolve them. So we decided the next thing to do was to use a smaller sample, revise and improve the paramagnetic spectrometer, and use lower microwave power so we'd have more sensitivity.

So the experimental setup was fairly simple, standard, with improved signal-to-noise. Because we poured in lower microwave power, we didn't have such hot electrons, and therefore the lines narrowed. We were the first ones to resolve the anisotropy of cyclotron resonance of electrons in germanium.

That was our first big success. The coauthors on that paper were myself, Zeiger, Dexter, and Rosenblum [P7.4]. Dexter and I were primarily responsible for the experiments. Rosenblum also helped work on the experiments. Zeiger and I did the theoretical interpretation, although the theoretical interpretation was simple. I just used the formula developed by Shockley [7.17] to fit the data. And that's all there was to it!

Dexter, however, didn't want to publish the data for a couple of weeks. The relative intensities that Shockley predicted [7.17] were proportional to the degeneracy of the particular resonances in a particular crystal direction. Dexter couldn't understand why the lightest mass, which was singly degenerate, was more intense than the higher mass along the one direction which was, in this case, I guess the 100 direction, which was 3-fold degenerate and should have been more intense.

Then it dawned on me a couple of weeks later that we were doing ionization. If the mass is small the electron gains much more energy for the same microwave power, and therefore you excite many more electrons from the impurity level at the lower mass than at the higher mass, and that explained the discrepancy. Then Dexter agreed to let us publish it.

So I was the one who made the key interpretations, and also had conceived of the experiment. I decided that on this one I should be the first author and properly so. We published this [P7.4], and that became a famous paper.

Optical excitation improves resolution

We couldn't do the ionization on the p-type. Then I got the idea from Burstein [7.23]: Why don't I try light excitation? I said, "That's a good idea." When I did that, with both infrared and white light, we got both holes and electrons. We were able to excite in a more controlled way, and we could use less microwaves, so the excitation and the resonance sampling narrowed the lines. Then we were able to resolve the anisotropy of the holes [P7.5].

I recognized that the anisotropy was analogous to the anisotropy that we had in magnetic resonance. I wrote an energy-momentum expression to a higher order which included the next term, using symmetry arguments, to get the next term in the anisotropy. But it was a fourth-order term and this didn't quite fit the data. It wasn't the proper thing.

But just about the time we were wrestling with the interpretation, Gene Dresselhaus and Kittel published [7.26] the results of the **k·p** calculation, which showed an expression which essentially is a quadratic in k for both the light hole and the heavy hole, which were essentially coupled by an off-diagonal term. Consequently, we used their expression, and Zeiger developed the Boltzmann transport theory, with my suggestion, in which we obtained an expression which fit the data. So we had that. We sent a copy to Kip and Kittel after it was accepted [P7.5].

Joint publication with Kittel's group on cyclotron resonance in n-type silicon

By that time we started doing n-type silicon, and I made the mistake of telling Kip that we had preliminary results. We weren't exactly sure but we thought it might be similar to germanium, but our experiments weren't yet complete. I did that on a Friday night. It was my mistake.

Over the weekend, Kittel and coworkers modified their apparatus and redid our experiments. But by the next week we already had the correct data. Kittel called me and he says, "It's along the 100." I said, "Yes, we got it." He says, "So have I." He compromised our data and so we decided, I being the generous sort of person, I said we'd publish jointly. It turned out our data was better.

So essentially, Dexter and I, together with Dresselhaus and Kip, published the n-type silicon [P7.6]. The p-type silicon Dexter, Zeiger, and I did, and we published by ourselves [P7.7]. That we didn't tell Kittel's group about.

So I would say, experimentally, Dexter and I were the ones who really measured all the parameters, with some theoretical help from Zeiger. This was the status by the fall of 1954. We had all the "Letters to the Editor" in *Physical Review* [P7.4–P7.8]. Ours were the only "Letters" that determined the anisotropy of the effective mass parameters of electrons and holes in germanium and silicon.

Invited paper at the Amsterdam conference

Then we were asked to present an invited paper on germanium and silicon at the Third International Conference on the Physics of Semiconductors in Amsterdam in the summer of 1954 [7.27]. Even though we had silicon data, Dexter didn't want to let us publish a paper on that because he wanted to reserve this for his thesis, and the rule at Wisconsin is you can't publish before you do your thesis. It was unfortunate.

But when I got to Amsterdam, I presented it for germanium and silicon [P7.8]. While I was there, Kittel asked me if I was going to publish these results in *Physical Review*, and naively I said, "Not until Dexter finishes his thesis." Which was supposedly going to be in the fall of 1954.

I do remember that was our first conference, our first invited paper at an international meeting, and of course we were the hit of the show. That was the time I met Hatoyama, and he was quite taken with me. We got to be very friendly. That's how I got Nishina to come to the Magnet Lab, because Hatoyama was the one who recommended to Nishina that he work with me.

This conference, the Amsterdam conference in 1954, was one of the most exciting conferences on semiconductors. Not only did we report our results, but Frank Herman [7.28] from RCA reported on his calculations of the band structure. Right before the conference he kept calling me, asking me for our data, so he could "fudge" his energy bands into the right place.

Frank Herman made claims that were not valid, but many of the results came from conversations with me, and he did give us credit for discussions. But he couldn't get the absolute energies right. However, from our data, we were able to identify where the bands were in germanium and silicon, and he was able to use that data to draw up the proper diagrams for the energy bands.

Cyclotron resonance successes required a team effort

There's no question Dexter deserved a great deal of credit for his experimental capability. But he essentially executed the things that I directed him to. As far as the interpretation, I did most of it, with help from Zeiger.

So it was a team effort. I mean, no one man deserves credit for the accomplishments and subsequent publications. I always put, certainly on the experimental side, Dexter's name first, except for the *Physica* paper [P7.8], which reported germanium, which I felt was my idea.

Extends cyclotron resonance to indium antimonide

This really got us into the semiconductor game, into the physics. We tried experiments on indium antimonide, Dexter and I. Unfortunately, we did not have a good sample of indium antimonide. I guess we got it from RCA. So did Kip and Kittel, and we both did the resonance. We didn't resolve ours. They resolved theirs. They saw the hole and the electron, but they didn't see the anisotropy of the hole, but they got the masses of the electron and hole [7.29]. We estimated the masses, but our experiment was not as clear cut as theirs. But then I started doing some theoretical work to try to interpret those results. This was I think 1954 or 1955.

Infrared cyclotron resonance

That's when I think I got the idea to go to the infrared [7.30], using pulsed high-field magnets. From my plasma physics data, I was familiar with Kapitza's work using high magnetic fields [7.31]. Apparently, Foner [7.13] was familiar with this technique, and I had suggested to Foner maybe we ought to leapfrog with indium antimonide … go to the infrared, use pulsed magnets. I think it was in 1956 when we decided to embark on that project.

Dexter and I did this experiment in indium antimonide in 1955. We gave a paper at the 1955 American Physical Society meeting [P7.9a]. We made an estimate of the mass to one significant figure, which was close, but not as good as the mass that was determined by the Berkeley group [7.29]. But nevertheless these two were the first experiments.

About the same time, we also, Dexter and I, decided to do semimetals [P7.10, P7.11], and similar experiments were being done about the same time, again in 1955, almost simultaneously by Galt [7.32] and company at Bell Telephone.

This began very interesting experimental and theoretical work on plasma effects in solids. Actually, we encountered the plasma effect in semiconductors in the indium antimonide work. The Berkeley group began to study this experimentally and theoretically. Subsequently, Laura Roth [7.33] and I began to study the theory for the plasma effect in aniso-tropic crystals like n-type germanium, and we developed this theory which she and I published in 1955 [P7.12]. I derived the susceptibilities, or con-ductivities, of germanium, including the plasma effect, taking into account depolarizing factors. We'll get into that later.

Infrared cyclotron resonance of indium antimonide with pulsed magnets

Let me get to indium antimonide. We decided that, by going to infrared, we'd avoid the plasma effect. It was well above the plasma frequency and would do a much better job.

We began our experiments. We divided the tasks. Zwerdling [7.34] and Keyes [7.35] were to build an infrared detector using zinc-doped germanium [7.36] to operate at low temperatures. Foner [7.13] and Kolm [7.37] had the job of building the pulsed magnet *a la* Kapitza. And I ordered a power supply, a condenser bank power supply, from Raytheon that was used to magnetize magnetrons as the pulse source for the pulse magnet.

The magnets that were developed were sort of the Bitter [7.38] type magnets, or at least similar to that, in which the helical coil was machined out of copper-beryllium. In between were put insulators similar to the way Bitter did it in his water-cooled magnets, and then these were clamped together. It made a monolithic strong structure. These magnets were cap-able of going up to, in a small bore, up to three quarters of a million gauss. But for the experiments we wanted to do, which had a larger bore of several millimeters, we went up to about 300 kilogauss. So that was the idea to do these experiments.

We did these experiments. The first time we tried was around Christmastime in 1955. We blew up the sample! So unfortunately, we didn't get any results.

Then in the spring of that year I met Burstein [7.23], and he was telling me he was doing these infrared cyclotron resonance experiments with dc magnets. I told him we tried it in pulsed magnets and were about to repeat it … we hadn't got any results. He was upset by it. We got our results about a week later, and I told him that we had the results at very high fields. Apparently this created a competitive and slightly antagonistic rift between the two of us. Yet, we had mutual respect for one another.

But the experiments really complemented each other, and there was no necessity for their feeling alarmed by our work. He thought we had done this work because he had done it first, but that wasn't the case. I told him we had actually tried the experiment but failed, even though we did it after he did. In other words, he was the first one to do infrared cyclotron resonance, but we had planned that experiment a year earlier.

But he didn't buy this. He thought I did it because I found out he was doing it. But this is the kind of misunderstanding that sometimes occurs. I think he later realized this wasn't the case. Certainly when the results came out, and I did the theoretical interpretation, it was clear that they got one point. We got a lot of points, and were able to explain it theoretically. In other words, again, ours was a much more complete work.

Nevertheless, this caused some friction between the two groups. In other words, they thought that we stole the idea from them. But that wasn't the case.

The idea of building a pulsed magnet occurred to both Foner and myself because we were familiar with the Kapitza work. I was familiar with Kapitza's work from my studies on fusion. We got the idea for that kind of magnet. I'm sure Foner and Kolm got it from Francis Bitter and it was their idea. I would not put my name on the magnet publications, and I never did. The authors were Foner and Kolm [7.39]. But actually I was the instigator.

First of all, I had done the analysis and I knew about the plasma effect. Why did I want to go to such high frequencies on n-type indium antimonide? You have to leapfrog the optical phonon. That's why we went to short wavelengths. And I knew the mass of indium antimonide from our experiment, done by Dexter and myself [P7.9a], and from the Kittel experiments [7.29]. Therefore, I knew what kind of wavelength was required. In fact, in anticipation of this experiment, long before we built the magnet, I ordered one of the first Perkin-Elmer infrared spectrometers.

So all the three parts, the spectrometer, the detector, the magnet, were being worked on. This took at least six months. This was truly in the old-fashioned way of developing all the components yourself. We knew enough about semiconductors to be able to identify zinc-doped germanium [7.36] as the logical detector, and that was the work of Zwerdling and Keyes. We essentially put Zwerdling's name first on that publication because I think this was the key element, although Foner and Kolm deserve a great deal of credit for building these beautiful magnets. That was an ingenious piece of engineering and development [P7.13, P7.14].

Cyclotron resonance in bismuth

In the meantime, Dexter and I continued our work on bismuth, and Laura Roth and I continued our work on plasma effects. The paper that Laura and I published [P7.12], which was I think my idea, but Laura did some of the detailed work, was to work out the magneto-plasma expressions for a complicated structure like germanium, n-type germanium, the anisotropic case. In fact, these expressions for the plasmas were redeveloped later when the electron-hole droplet was discovered [7.40], but again nobody referred to our publication, which preceded this work.

In connection with bismuth, I also wrote a paper on magneto-plasma effects [P7.11], in which I derived the expressions for what, later, were called by Buchsbaum [7.41] the lower and upper hybrids, which involve both holes and electrons, and which are the equivalent in a gaseous plasma to ions and electrons. I derived these expressions long before he and Allis [5.9] categorized them. But, again, no one referred to this paper, even though later, when Buchsbaum talked about it at a presentation at the MIT RLE, I told him, "All this is in my paper on bismuth, but in a more generalized form."

Of course, I derived the expressions for cold plasmas, but later I'll talk about work which we did later, in which I did a quantum treatment of these, and which works beautifully for hot plasmas too. But this takes us aside.

The point is this. By working on indium antimonide and bismuth, I was laying the foundations for subsequent work that was rather important. In fact, in the data on both bismuth and indium antimonide that Keyes took [P7.13], he kept telling me that there was structure in the resonance spectra, which we thought was cyclotron resonance. They were low mass materials, so they were ideal.

Anomalous skin effect in a magnetic field

However, I suspected that it was the anomalous skin effect in a magnetic field. So, I started deriving the Boltzmann transport theory of the anomalous skin effect in magnetic fields. To my knowledge no one had done it as yet.

I wrote the integral differential equations, except I didn't know how to solve them. I asked Neustadter [7.11] to do it for me. He kept arguing with me because, when you do the anomalous skin effect, you have to take into account specular and diffuse reflection from the surface. But you could do it for one or the other, and you would get the physics out of it. That's what I wanted him to do. I wanted him to study the specular reflection, which was simpler, but he refused. So we never got it done. This was around 1956–1957.

In the meantime, two Russians, Azbel' and Kaner [7.42], attacked the same problem, I think maybe even after I did. In fact, you can look at the 1956–1957 progress reports, and equations that I derived are in there. But they completed it and discovered the Azbel'–Kaner effect, this anomalous skin effect resonance which was later observed by a number of other people like Kip and coworkers and a guy in Canada. So this was the anomalous skin effect resonance.

Now this problem was very similar to the problem that Eugene Gross [6.8] tried to solve in the magneto-plasma effect, magneto-plasma waves in a gaseous plasma, in an ionized plasma. I mean primarily electrons, because both of them were the solution of a Boltzmann equation and it involved integral differential equations. You've got Maxwell's equations and you substitute an integral expression for the conductivity in the integral which involves the electric field. So it's a complicated problem.

But the way to solve it is to expand it in a Fourier series, harmonics of the cyclotron frequency, something which Neustadter should have known. But I didn't know this was the way to solve these integral differential equations. Had I studied advanced calculus in college, I probably would have known. Later I learned the solution.

So these were very key problems and they're the basis for a lot of these interesting phenomena. I identified the physics in both cases. Again, I was ahead of my time. It was too bad that I didn't have the right theoretical help. It seemed to have been my good fortune to be on top of advanced problems each time, often ahead of the rest of the people.

Nevertheless, I took a different approach. I decided to teach myself a lot more about the theory of the optical properties of semiconductors. I was the one who did the quantum theory and interpreted the cyclotron reson-ance data in indium antimonide. I was the one who published the paper in 1961 [P7.15]. Even though I had coauthors, I really did most of the theory. I had some help from Mavroides [7.43] doing the numerical work, but essentially it was my theoretical work.

Solves Klein–Gordon equation

The interesting story about that is one day I was driving down Route 128, going home, and all of a sudden I was solving the relativistic quantum equation in my head while I was driving. It was essentially the Klein–Gordon equation.

But I did it a little differently. I took the Dirac equation and expanded the momentum p as a series, inside the square root, as a series in the momentum p: p^2, p^4, and so on. And I said, let the p^2 term be the dominant term and let the others be perturbations, and I then derived the expression. I said, "O my god, I could put this back under the square root." And then, essentially, I solved the quantum mechanical problem, which is the equiva-lent of the Klein–Gordon equation.

I'm not even sure at that time I had heard of it, Klein–Gordon, but I started reading up on the literature. So I came back a couple of days later and told Laura Roth [7.33], "Look I've solved this problem." So she said, "That's interesting. A couple of people at Westinghouse are also attacking this problem slightly differently." This is indium antimonide. But I said, for me the relativistic equation, Dirac equation, was the interesting thing. So I had solved it in my head. I've done this a number of times.

Like Laura Roth used to say, "You write out the answers, and it takes me three weeks to find out if you're right or wrong."

For example, I derived the equation for the bismuth and the lead tel-luride anisotropic bands in a magnetic field, the ellipsoidal bands which are non-parabolic, and Millie Dresselhaus [7.44], who later did a lot of work on bismuth, often gives me credit and calls it the Lax Model.

This is very similar to the work without a magnetic field that Morrel Cohen [7.45] and one of his associates did. I don't know whether I got the inspiration from him. Although he claims I gave him the inspiration for the work that he scooped, that Laura Roth and I were doing on bismuth. I wrote out the energy bands for bismuth, and Laura was given the task to justify my expression using group theory. Before she could complete it, Morrel Cohen published the paper and gave me credit for the inspiration because I wrote out the expression when I gave the Buckley Prize talk in 1960 [7.46].

Now, just thinking about it, I figured out what the band structure of bismuth would look like because it was both anisotropic and non-parabolic. It was ellipsoidal, but I included the non-parabolicity. Morrel Cohen [7.45] wrote a paper which justified my result, and he gave me an acknowledgment, and apparently Peter Wolff [7.46a, 7.47] also did the same thing.

But the point is, from the indium antimonide work and the bismuth work, I got the idea, from the quantum treatment, of doing the interband experiments.

7.3 Interband Magneto-Optics

Discovers interband magneto-optics in germanium in 1956

This period of 1956–1957 was a very exciting period for me. Even though I was a group leader now, in the evening I would work at home, and constantly I would be solving problems whenever. I was preoccupied with all these exciting phenomena of cyclotron resonance in the quantum limit and magneto-optical phenomena. My idea was that we would do germanium.

I read the GE paper [7.48], in 1955 I think, and I decided to have Zwerdling do a new setup. We ordered a spectrometer and another magnet, and he was to do interband absorption in germanium.

I got the germanium sample from Dash and Newman [7.48]. These people polished a crystal of germanium 10 microns thick, which was too thin to see the indirect transition. Because the indirect transition is a weaker transition, you need a sample several millimeters to a centimeter thick to see its features. But if you make it very thin you don't see it. So you see the direct transition in germanium, and I recognized that this was the experiment to do.

We set up and did this experiment in, I think, the spring of 1956. We got the apparatus together and we did it at room temperature. We observed this oscillatory effect, a new phenomenon. Gee, I was excited!

Kittel came to visit and I showed it to him. By that time we were on good terms again. I showed this to him and he said, "Gee, this is an important discovery."

I wanted to publish this, but Zwerdling, in his typical way, cautioned that we had to do it more carefully, and we were planning to do it at low temperatures. But even at room temperature we saw this oscillatory effect. I interpreted it for what it was, and estimated the effective mass of the upper conduction band, which checked very well with Gene Dresselhaus' estimate [7.49] from the **k·p** theory using the cyclotron resonance parameters of the valence band, because the valence and the conduction bands were sort of coupled if you did the **k·p** theory in its full-fledged fashion. That was amazing.

In fact, this value and this phenomenon are referred to in Kittel's advanced solid state textbook. He doesn't refer to my cyclotron resonance work, but he refers to this work.

That was the first time that the mass of the gamma point electron was experimentally determined in germanium. It was an important discovery. This was done in 1956.

At the same time we began to build a room down in the basement of Building C for Zwerdling to do these kinds of experiments at low temperatures. We recognized that this would be a very important new technique for studying semiconductors. I think this was done in the spring of 1957.

In the fall of 1956, we had finally submitted this for publication [P7.16]. I went to a conference in October 1956 in Washington, DC [7.50], before it was published. At the conference, Burstein and I independently reported this phenomenon. He had observed it in indium antimonide [7.51], I had observed it in germanium [P7.16]. But apparently, we were much further along, both in the experiment and theory.

Laura Roth joins in 1957

By that time, we had the low temperature apparatus, and Laura Roth [7.33] joined me full time, I think that fall, in doing these experiments. So, we did experiments in indium antimonide as well as germanium [P7.17].

As I looked at the indium antimonide data, I recognized the splitting of the conduction band. I interpreted it as a splitting due to the spin, and we were the first ones to actually experimentally measure the anomalous g-factor in indium antimonide [P7.17].

This was, I think, September of 1957, or summer of 1957, but shortly after Laura Roth joined Lincoln full time, I gave her Luttinger's paper [3.4, 7.52] and told her she should be able to interpret this thing in terms of the Luttinger **k·p** model in which he took the off-diagonal terms as calculating the g-factor. She did that and derived her formula for the anomalous g-factor in indium antimonide. It checked very well with the experiment, and she became famous for this [P7.18, P7.19].

Laura and I subsequently published a long paper with Zwerdling [P7.19] that contained this result and her theory. She became famous overnight for this work. But actually I was the one who, let's say, gave her the problem. I recognized what had to be done, and I thought this would be a good first postdoctoral project for her. It turned out to be the case.

Of course, because I had worked with her in previous summers, I knew she was bright and that she could do these things. That's why I hired her. It turned out that she was a very good member of the team of Zwerdling, Lax, and Roth, and we did a lot of work together.

Detects exciton in germanium

Laura ultimately interpreted the low temperature work, which had a lot more fine structure, in which we observed the exciton too. Zwerdling and I interpreted the binding energy of the exciton, and measured the energy gap fairly accurately [P7.20].

The thing we didn't realize is that there was additional structure due to strain in the sample, which we did not anticipate. The sample was mounted on a sort of window, and when you went down to low temperature the frame had a different coefficient of expansion than the thin sample, and it would strain it.

So Laura identified most of the important transitions, but couldn't account for any splitting or structure due to the strain. These are things people today are doing much better. But Zwerdling and I had enough resolution to do this. The original experiments were done with a double-prism Perkin-Elmer spectrometer, I think one of the first they sold.

We were among the first to do this, and we were the first to actually study the optical properties of semiconductors here at Lincoln. Then we bought the grating spectrometer to resolve the structure, which ultimately Laura Roth identified. So this was done I think in 1957 or 1958.

We were the first ones to observe the direct exciton in germanium [P7.20]. The British group at Great Malvern had observed the exciton in the indirect transition in germanium and they interpreted it, but they never observed the direct exciton because they never had a thin sample. So we were the first to see the direct exciton in germanium, and we calculated its binding energy. I don't know if we saw the effect of a magnetic field. But anyway we got the exciton.

Then we attacked the indirect transition with a thick sample [P7.21]. Here, the indirect transition in a magnetic field was a step function, as Laura and I derived the theory that showed it. Laura did it quantum mechanically, whereas I started developing these things in a semiclassical fashion using the Kramers-Heisenberg expressions, which is a quantum mechanical expression. But then the rest goes essentially phenomenologically. We both got the same result. It's the step function.

So the Landau transitions were a step function, and so was the exciton, to my surprise. But we knew the difference because we plotted in the "fan chart" that I had invented to interpret the first experiments on these magneto-optical effects. [A "fan chart" is a plot of the transition energies versus magnetic field strength. Transition energies are determined from the experimental data for transmission or reflection plotted versus photon energy.] I asked Zwerdling to plot the results, energy versus magnetic field, for these various lines that we saw, and this enabled us to distinguish between the Landau transitions and the exciton.

Friendly rivalry with Burstein's group at the Naval Research Laboratory

Both Burstein [7.23] and I were doing similar experiments. But, between the theory that Laura and I were engaged in – mostly Laura but with some input from me – and some of the phenomenological theory that I was doing, and the experiments that Zwerdling was doing, we surpassed and did a lot more than the NRL group that was doing similar experiments.

At any rate, we both were on the program committee [for the 1958 Semiconductor Conference in Rochester] and decided both groups would give invited papers [7.53]. There was an argument as to who should give the first paper, Burstein or me, and apparently for some reason it was believed that his was the more influential, and I was to present second. Nevertheless, when you look at the proceedings [7.53], our work was the more definitive work. But similar. Anyway, our combination of experimental and theoretical work was more complete. We did both germanium and indium antimonide, and the interpretation by Laura was more complete.

We had a very good team. That was the beginning of a whole host of magneto-optical effects, and these, together with cyclotron resonance, are what won me the Buckley Prize [7.46] in 1960.

Interband magneto-optics of bismuth

Just about that time I got my Buckley Prize, in 1960, just before the Magnet Lab opened on July 1, 1960, I had given my paper in May of that year at Johns Hopkins, and I talked about cyclotron resonance in bismuth. Two

FIGURE 7.1   Dr. Benjamin Lax receiving the Oliver E. Buckley Solid State Physics Prize from Prof. George E. Uhlenbeck, President of the American Physical Society, at the annual dinner of the American Physical Society and the American Association of Physics Teachers in New York City on January 29, 1960. (Photo credit: Eli Aaron. Photo use courtesy of American Institute of Physics Emilio Segrè Visual Archives, Physics Today Collection.)

weeks later, I realized something about the experiments that Bob Keyes had done [P7.13] with a pulsed magnet. He had observed structure, which I thought at one time was the anomalous skin effect. But now, as I tried to fit the theory, I decided it was an interband transition between a valence band occurring below the Fermi level, since bismuth is a semimetal, up to the conduction band, and therefore was very similar to what we had seen three years earlier in semiconductors. In fact, we had seen this phenomenon before we had done semiconductors, but didn't realize what it was. Then I realized, you didn't need a pulsed magnet to see this. We could do it with a dc magnet.

Once I realized it, I began a set of experiments with Mavroides and Dick Brown. I set up Mavroides and Dick Brown with a Varian magnet and a low temperature dewar, and we did a reflection experiment. We observed the interband transition, just as I suspected [P7.22, P7.23].

Brown had done a thesis [T54] with me on Faraday rotation, where we had discovered interband Faraday rotation. He observed the phenomenon and he couldn't explain it. I realized what it was, and we started theoretical and experimental work on the interband Faraday rotation. This began to be the postdoctoral project of Nishina, who joined us in the Armory.

Millie and Gene Dresselhaus join Lincoln Laboratory in 1960

Just about that time, after we discovered this, Millie Dresselhaus [7.44] came for an interview at Lincoln. I showed this to her and she said, "This is what I want to do next. This is what I'd like to do for my research."

We made her an offer. IBM made a similar offer. I think IBM's was a little bit better than ours, but she decided to come with us.

Shortly afterward, she joined our effort and worked with Mavroides and Dick Brown on this phenomenon. We published the first paper [P7.23] on this in *Physical Review Letters*. I put her name on it because she helped to reduce the data and I figured, you know, it doesn't hurt.

These two papers [P7.22, P7.23] definitely established the interband magneto-optical phenomenon in semimetals. This actually started Millie Dresselhaus' career. She decided to devote herself to this project, and she teamed up with Mavroides after Dick Brown left Lincoln for the Magnet Lab at MIT.

Millie Dresselhaus takes over magneto-optics in bismuth, graphite

When I left Lincoln in 1965, this work was taken over by Millie Dresselhaus together with John Mavroides, and they pushed it to a fine art. They did some very important work, particularly in graphite, for which she became well known and well established as a first-class scientist.

Of course she had the help of her husband, Gene Dresselhaus, on the theoretical side. This made a powerful team for this kind of work, for studying both the energy bands and the Fermi surfaces in some of these semimetals like bismuth, arsenic, graphite, bismuth-antimony alloys, which pushed it into a different class of materials, some of which were semimetals and some of which, with the proper alloying, became semiconductors. So it was a very interesting study. Here is where, again, the high magnetic fields played a role.

Six-week grand tour: Great Malvern, Brussels, Paris, Grenoble, Zurich, Eindhoven

In the summer of 1958, around the time of the semiconductor conference [7.53], I took a trip to Europe, with Zwerdling joining me for the first part of the trip.

We first went to a conference [7.53a] at Great Malvern. The reason we went was that we were invited to participate in a conference in Brussels. It was a World's Fair [7.54]. And we were asked. Shockley was to be there, Bardeen, Esaki, who had discovered the tunnel diode, and myself, all to give invited papers at this conference.

I decided to make it a grand trip, and take Blossom with me, to go to Great Malvern to another conference and present some of our results, which of course were complementary to the work done by the Great Malvern people on the indirect transition without a magnetic field. R. A. Smith [7.55], head of the group at Great Malvern, was the one who invited me.

So I was to go to London, then to Belgium for this conference, and then we decided we'd make a longer trip of it and I was invited, or maybe I finagled invitations, to Paris by Aigrain [7.56] and his group to give talks at the *Ecole Normale*.

Then I was supposed to give a talk in Grenoble, where I guess they hadn't yet planned a magnet lab, but I don't know whether they were thinking about it. They may have had a small magnetic field facility, but not the ultimate national lab.

Then we were to go to Switzerland, and in Zurich I was scheduled to give a paper. Then on through Germany, and onto Holland and the Philips Laboratory, where I was supposed to give a paper. So it was a grand trip, about six weeks.

Great Malvern, England

Great Malvern was great. Zwerdling and I both presented papers at Great Malvern. We had a marvelous time there. There was a banquet. I was asked to speak for the American contingent as a representative. R. A. Smith and his people were wonderful hosts. He and his wife were very gracious. We became very good friends with them.

From there on we went on to Belgium. We started driving. We went to Paris … drove through it. We stopped in Lyon, then we got on to Grenoble where I gave a talk. Then we went on to Chamonix.

At Chamonix, who do I meet but Shockley. He was skiing down the hill. At the end of June, there was snow up on top and we went all the way up to the top and on the way I met Shockley and his wife and we became good friends. Of course we were at the Brussels conference together with Bardeen and his wife, and so it was a congenial group.

When I went to the top of the Chamonix there was a big boulder, an enormous boulder about eight feet high which was at the peak. I insisted I had to get up to the peak, had to climb up on this. I didn't realize how high we were. After I got up I was almost out of breath, but I ended up on the top of Chamonix. That's when I began to appreciate what the altitude does to you.

We came to Zurich. We got to Geneva. At that time Weiskopf [7.56] was the head of Geneva CERN and he had been in an accident. Blossom and I went to visit him. We paid our respects. I think we had had him at MIT. We wished him a speedy recovery. Later, I got to be one of his colleagues, but at that time I thought it was the right thing to do.

In Zurich, we went to the *Technische Hochschule*. There was a well-known physicist who was the head of it. I forget the guy's name. I gave a talk there and it was well received. In fact all my talks were well received.

We visited Lucerne. We visited all the great cities in Switzerland and it was beautiful. From Geneva we went on to Germany, to the Rhine Valley. We drove along the Rhine. It was raining. As we drove along the Rhine there was a big barn with the name "Professor Lax" on it. Yes! We didn't have time to stop but that was interesting. I don't know, he must have had a school or something there. We went on to Cologne.

It was my chance to visit with Polder [7.57] and company, and with H. J. G. Meyer [7.58] who I got to know at the Amsterdam conference in 1954. We got to be very good friends.

And it was there that I got the diamond pieces for Blossom's ring that I eventually designed for her and had made here in America.

So we met old friends, Polder [7.57], H. J. G. Meyer [7.58], and the others, Gorter [7.59], and we were very well received and I gave a number of talks. In 1958, they were very interested in what we were doing because the Philips people were very much interested in semiconductors as part of their research work, also in ferrites, and I also was working in that field too.

I don't remember whether I gave talks in both, but I probably did. We had a lot of common interests.

We visited a couple of days at the Philips Laboratories and I met the Director, Casimir [7.60], whom I had met before. In fact, we once had breakfast together and he always mixed me up with Melvin Lax [3.2]. [*Melvin Lax was a well-known theoretical physicist, a contemporary of Ben's, but no relation.*] I never straightened him out.

But anyway, I was known to him. So we had a good relationship and very good rapport with Philips Laboratories. We had lots in common, both on ferrites and on semiconductors, and it was a leading laboratory on the continent in these fields. They were primarily interested in pushing transistors and the use of microwaves, ferrites and microwaves and computers. Remember, at the time, memory core was the heart of the computer. This was the work that was begun here at MIT and Lincoln Laboratory by Forrester [7.61].

But the work on ferrites that gave rise to both the microwave and the memory core was pioneered at Philips. Gorter [7.59] and his people were the ones who developed those things.

Meets Néel in Grenoble

The reason I went to Grenoble was because I wanted to meet Néel [7.62]. At that time, he didn't have the Nobel Prize yet. He was the father of the antiferromagnetic and ferrimagnetic structure and its magnetic properties. In fact, the critical temperature is not called the Curie temperature, but is called the Néel temperature.

I was in the process of writing a book at the time [P7.24] so I knew about Néel's work and I met him. He was a rather formal gentleman and a slightly older man, so he treated me more like an elder statesman rather than as an equal. But I met him. Ultimately, he got the Nobel Prize for this work.

But I recognized the significance and importance of this work because I was in the process of writing the book. In fact, I was probably at the stage where I was writing about his theories on the structures. So this was a very successful trip.

A worthwhile trip

I combined traveling with pleasure, but the idea was to meet these people, these leaders of the European physics contingent, in England at Great Malvern, with whom we had a lot in common, and in France with Aigrain who was working in semiconductors. All of these people who had gotten to know me for the first time in Amsterdam [7.27] were delighted to receive me there, including Néel and his people, and his disciples at Grenoble and Zurich, and at the *Technische Hochschule* where there was some critical work going on, and, of course, most importantly, the group at Eindhoven. So I learned a lot too.

So this was both a pleasure trip and a business trip. It was sort of a busman's holiday. It was intellectually very stimulating and of course I came home with a lot of exciting ideas. This is what happens when you attend these conferences.

Back to writing a book with Ken Button

So here I was in 1958, having had success from the cyclotron resonance and magneto-optical work. I was in the midst of writing my book with Ken Button. I think in 1958 we had just the preliminary notes, including the perturbation theory and all those things that I had developed here in the lab, and some of the devices we were working on, explaining the device of Gerry Heller [4.17] on the non-reciprocal isolator. In fact I used the perturbation theory to explain it. He was chagrined. He never forgave me for that.

7.4 Masers and Diode Lasers

Maser work at Lincoln Laboratory around 1956

In the 1950s, I encouraged people to do thesis research. One thesis [T59] was done by Dick Dexter [7.16] on cyclotron resonance. He was from the University of Wisconsin, and so was Jim Meyer [7.15]. Jim Meyer wanted to do a thesis on paramagnetic resonance, on his own time. He would bootleg on much of the apparatus that we had for magnetic measurements, using all the big magnets that we had. In fact, Jim built a magnet for himself, very similar to the Varian magnet. He used this magnet, I think, to do his paramagnetic resonance.

Just about that time, Bloembergen [7.63] had invented his three-level maser. He was a consultant to us at Lincoln. In fact, he got the idea of the three-level maser from an idea that Zeiger had discussed with him for a three-level laser. But it turned out that the idea used a gas, which I don't think worked, but it was certainly one of the first to propose the three-level pumping concept.

Bloembergen decided that the place to do such a thing would be at microwaves: use one microwave frequency and pump at a higher energy, and come out at a lower energy to make a maser. He worked out a paramagnetic approach: do it in a paramagnetic material. He in fact proposed as one of the candidates the material that Meyer was using for his thesis – that he had used – I think he had completed the thesis.

I decided that this was a very worthwhile project, and I asked Al McWhorter [7.64] to drop his work on surfaces, which I didn't think was that productive. I had taken over the large group by this time. I suggested that he and Meyer team up, using Meyer's apparatus to build a three-level maser. He and Jim Meyer, I think McWhorter particularly, began to study various possibilities, and they came up with a better crystal, potassium chromicyanide, as the best candidate for their maser.

They asked Harry Gatos [7.65], who became the Head of the Materials Group at Lincoln, to grow such a crystal, and this was the first chromium maser ever.

In the meantime, I think before they got theirs working, the Bell Labs folks had built one. They picked another material and built a three-level maser according to the concepts developed by Bloembergen. But the first chromium maser was this chromicyanide maser of ours, and McWhorter and company deserve the credit for it.

It worked very well as a low temperature amplifier. We kept working on the maser projects. One of the ideas that came up, suggested by Stan Autler [7.66], was this: since these experiments were done at low temperature, because they wanted to make a low temperature amplifier-detector at liquid helium temperature, it would be a good idea to make a superconducting coil instead of using a large, external Varian magnet to use, let's say, in a practical maser. So, he wound a niobium superconducting magnet around one of these coils and indeed it worked, it worked very well. And this was the beginning of the high field superconducting work.

I think this occurred around in the late 1950s or early 1960s. We had a visit from Bell Telephone Laboratory. They presented some of their work and we presented our work. When we presented our work on the superconducting magnet, Kompfner [7.67], who was the head of the research group at the time, says, "How come we're not working on it?"

They went back and really began to work on high field superconductors, which ultimately led to the niobium-titanium and niobium-tin work, but I think it was the work of Stan Autler that triggered their activity. This was one of the projects that we decided to make an important part of our high field program.

In 1961, we planned a high magnetic field conference [7.68] to talk about our development of high field magnets, the Bitter magnets, and we included a session on superconducting magnets, which were being developed by the Bell people. This was one of the big highlights of this conference.

Quantum Electronics Conference in 1959, request from Charles Townes

About that time, Charlie Townes [7.69] asked me, I think about 1957–1958, to join a committee that was going to have the first quantum electronics conference [7.70], and asked me to prepare a paper on the possibility of a semiconductor laser.

So I began working on some of the theoretical aspects. Herb Zeiger and Stanley Autler were also working on the possibility. We were thinking mostly of intraband lasers, and myself on the cyclotron resonance optical maser as we called it. Zeiger and Autler were thinking of an impurity transition, very similar to the three-level maser that Bloembergen proposed. In other words, pump to higher energy, and then invert lower energies, essentially a three-level type laser.

But some of us subsequently began thinking about an interband laser. I think Herb was thinking about it. I was thinking about it in terms of magneto-optics, which ultimately I called the magneto-optical laser.

The conference was held [7.70]. I presented a paper [P7.25] on the cyclotron resonance maser, and then briefly discussed the possibility of the interband laser. In fact, somebody asked me the question, it may have been Eli Burstein, and I came to the conclusion then, at that time, that the interband laser was much more feasible than the intraband laser, the cyclotron resonance laser that I proposed, although ultimately, some many years later, such an optical laser was built by the Russians. They developed a clever scheme of inverting, whereas I was proposing optical pumping, which may or may not have worked.

In any event, I came to the conclusion at that time that I wasn't going to work on the cyclotron resonance maser, that the optical maser was more feasible.

Has misgivings about surface physics program

The one program that I had misgivings about was the surface program that Kingston [7.71], Mino Green, and later, McWhorter were engaged in. Both from the experimental and conceptual side, I thought it wasn't on sound footing. The findings and objectives didn't appear to me to be too well defined. The first chance I had, rather than cut it off, I tried to divert people to other programs. That's what I did with McWhorter.

I felt he'd make a much better contribution with the maser, and it turned out that I was right. He was the key man that made that program possible, and that was a big feather in our cap.

Kingston jumped on the bandwagon later and began to build a maser to put on a radar, which ultimately I think saw a signal reflected from Venus. So for the first time, I think we made a contribution to a particular system, which I guess pleased many of the people at the laboratory, including the directors.

At the same time, this is what got Autler started, that same kind of project. He thought it was a good idea to put the superconductor on the maser rather than an electromagnet, and this is how the high field superconducting work started in this country, which was later picked up by the Bell people who visited us. Unfortunately, we could not compete with them. They had more money and more resources.

Autler wasn't the kind of man who was willing to share the program with others. He wanted to keep it to himself, which I thought was a mistake, but I concurred. In fact, his statement was, "I don't want you to work with me, Ben Lax. You're too fast for me." In other words, you know, I'd get ideas. So I stayed away. I felt it was his baby. I thought it was a nice project. I think it was one of the important contributions that Lincoln made with technology.

Magneto-tunneling in InSb

Speaking of tunneling in a magnetic field, actually I got the idea long before tunneling in superconductors was discovered.

Esaki discovered tunneling in the tunnel diode [7.72], and he gave a talk here at Lincoln sometime in the late 1950s. He wanted to study the anisotropy of tunneling in germanium as a function of crystal orientation. And I said to him, a much better experiment would be to put it in a magnetic field. That was much more sensitive. It sensed the nature of the effective masses as a function of orientation, as we demonstrated many years before in cyclotron resonance [P7.4]. Although I didn't really have germanium in mind for the tunneling experiment.

When we got out of the meeting, after my suggesting that, Rediker [7.73] and Kingston [7.71] sort of ridiculed me. He says, "You want to put a magnetic field on everything."

I said I had a very good reason. They laughed, and I said to Rediker, "You ought to do this experiment." He didn't do it.

Then, a couple of months later, I got a call from Esaki, and he told me he had done it in germanium in a magnetic field and he observed a small effect. I didn't say anything because that's not what I had in mind.

So I came and told Rediker, "See, Esaki was much smarter than you. He took my hint. But you ought to do the tunneling in indium antimonide."

I wrote out the answer for the tunneling probability in a magnetic field just from my mind, because I had worked with these things. I asked McWhorter to work with me on the formal theory using the WKB method. We did that and we got the same result. My answer was correct. I just merely took the non-magnetic formula, and by intuition and having worked with this sort of thing, was able to write the answer, including spin. And we formally showed that that was the correct thing.

Sure enough, when he and Calawa did the experiment, the theory and experiment agreed and we published the paper [P7.26]. So this was another kind of experiment that is a very good one in magnetic fields. We didn't pursue it at the Magnet Lab, although that would have been a good project in other semiconductors.

GaAs junction laser diode demonstrated at Lincoln in 1962

Well, Rediker doesn't agree with me, but Zeiger and I, and Autler I think, were working on semiconductor laser concepts. I reported this at the 1959 Quantum Electronics meeting in Shawanga Lodge in New York, and I gave Zeiger credit for his work [P7.25].

We listened to Aigrain at one time, I think it was around 1960 or thereabouts. He gave a talk, and he talked about making a semiconductor laser in germanium by using a Weierstrauss sphere and a point contact. Zeiger and I came out and we said, "This is ridiculous, it's an indirect transition. It has a very small absorption coefficient, therefore would have a small gain." I didn't calculate anything, and Zeiger ultimately may have done so. But I knew this qualitatively from having done these magneto-optical experiments and optical studies.

At any rate, when I gave my talk [P7.25], I concentrated on the cyclotron resonance maser, which was my pet project. But I did say at the meeting that that was an *intraband* laser, a weak laser, much more difficult, and that more likely the *interband* laser would be the way to go.

So in 1961, at a meeting with Rediker and Keyes – they may deny it – I said to them, "We've been fussing around with concepts and theories. It's time to work on a semiconductor laser." I proposed indium antimonide, optically excited in a magnetic field. Keyes apparently had been thinking about it, and although this wasn't unique with him, a number of people had suggested junction lasers. He thought this would be the better project to work on.

They started with indium antimonide. Indium antimonide wasn't good enough in those days to make a good junction, and they didn't succeed. But they wouldn't accept my idea.

So ultimately, they went to gallium arsenide. And, unknownst to us, a demonstration of the phenomenon, the phenomenon that Keyes and Quist discovered [7.74], was preceded much earlier by Braunstein [7.75] who took a point-contact gallium arsenide diode and showed that, when you took it from room temperature down to liquid nitrogen temperature, the emission increased drastically.

Keyes and Quist [7.74] did the same experiment with a junction diode, and there, of course, it was much more dramatic. You couldn't make a laser out of a point-contact diode. But a junction diode that covered a certain area and had certain dimensions, large enough for a gain path, was a much better prospect. And this was achieved, I think, about May of 1962.

Townes visited us, and I told him that, in about a month or so, we're going to have a laser. But Keyes and Rediker got off on a tangent and didn't attend to it. Zeiger I think was trying to get some diodes polished. In any event, these efforts were, to be frank, a little amateurish and they were not done well. Nothing came of it.

I began to worry about being scooped. In fact, we heard through the grapevine that IBM was working on it. I didn't know GE was working on it too.

So, anyway, I think we found out that GE had succeeded, and I said to Rediker, "It's time we got an effort." I called a meeting where we assigned different people to different tasks. We got somebody to work on the electronics, somebody to work on the low temperature system, and Rediker and company to make the diode laser. We decided to work both at liquid nitrogen and liquid helium temperatures.

But we were too late. GE published well ahead of us [7.76]. IBM beat us by a month [7.77]. But we came out [P7.27], I think, a month later than GE and IBM.

Nevertheless, it was the Keyes and Quist work [7.74] that demonstrated that the semiconductor laser was possible. I credit them with this seminal work. It was question of building the right Fabry-Perot structure and the proper electronics.

But this began the laser work here at Lincoln, and it was truly a group effort. Zeiger, McWhorter, and I did the more formal theory, primarily

FIGURE 7.2   Dr. Benjamin Lax. Photo taken in 1960. (Courtesy of the MIT Museum.)

McWhorter. I sort of asked him to work on the theory, and we consulted with one another, but he did the formal theory, which involved the transition probability, the density of states, and the statistics. He calculated the threshold, the gain on the threshold *versus* the loss [P7.28].

In a much less formal way, a calculation had been done by Dumke [7.78], which just simply used the inverse of the absorption coefficient and derived the formula for the gain needed to overcome the free carrier loss and the loss at the interface of the Fabry-Perot, which is a reflection loss. Once the gain exceeded that, you achieved threshold.

The theory of McWhorter was a much more formal and much more precise theory, but it was equivalent. Since then, more sophisticated work has followed, but the key work was the McWhorter theory [P7.28].

This was the kind of work we were doing at Lincoln in the early 1960s, just before the Magnet Lab opened.

Lincoln Laboratory *versus* the Magnet Laboratory

The two are intertwined. Until 1965, I was directing both the Solid State Division at Lincoln and the Magnet Lab. So we had a good relation. In fact, that was my intention for creating the Magnet Lab, to be part of the solid state effort at Lincoln, because I felt just having a lab that just is a facility is not as good as if it were part of a larger effort where you provided material, and where there were scientists who could take advantage of the Magnet Lab.

Ben's role in initiating programs

Both at Lincoln and at the Magnet Lab, one of my roles was to initiate projects. Sometimes they were my own ideas, and sometimes they were ideas that came from outside or occasionally inside in the laboratory.

For example, I'm the one who initiated the maser project, appointing McWhorter and Meyer as the ones responsible. I, of course, started the cyclotron resonance projects, the magneto-optical studies at Lincoln, the ferrite program, and the various subprojects under the ferrite program with Ken Button, and even experiments that Tannenwald and Artman did.

And then, of course, I approved some of the other projects in materials that were initiated by people like Harman, Strauss, and Gatos, and they would tell me they were doing this because it was their specialty. I would weigh it in terms of how it would fit into the overall program.

The only thing that was missing – and we did have some problems and criticism from the rest of the lab – was that our ferrite work didn't fit into the systems programs, although the ferrite work indirectly did relate to them. In fact, the work that Button and I did, particularly the invention of double-slab ferrite phase shifter, became the standard for the phase shifters of phased arrays, particularly those developed at Raytheon. Later on, I found out they used my theoretical work to do their calculations.

So, in essence, although our ferrite work may not have contributed directly to Lincoln, it did contribute to the advancement of ideas and technology, and it was picked up by others elsewhere. So, in that respect I think the Lincoln Solid State Group made some practical contributions.

Czechoslovakia conference

Another interesting trip was the one I made to Czechoslovakia. That was 1960. It was the conference on semiconductors. I gave an invited paper there [P7.29].

The contrast between the Czechs and the Hungarians was tremendous. Everybody was so serious and solemn in Czechoslovakia. When you got there you knew the communists were in strong control. The military would inspect you for everything. When I was there, they apparently rifled through my bags but didn't take anything.

I had taken lots of photographs in Czechoslovakia. I took the pictures of the oldest synagogue there, the one with the cemetery next to it where they piled people on top of one another because the Jews never got enough land to have a good cemetery. There was a Holocaust Memorial there. They had a clock in Hebrew which went the other way. I took a picture of that. Apparently they opened up the camera and destroyed the film. I never had it developed because the film was exposed. So they came in there to do that.

Whereas, when I went to Hungary, which I think may have been a year later, the contrast was so great. The Hungarians were a little bit more relaxed. Communism wasn't taken as seriously there, and people seemed much more cheerful and brighter and friendly. So I liked that.

There's a story about Harry Gatos and myself at that conference. In order to buy anything in Czechoslovakia, you have to change dollars to their currency. There was a department store, and downstairs was where you went to change currency. As Harry Gatos and I were waiting in line, a couple of young women, I guess in their early twenties, came up against us and began to put their fingers down our back, you know, trying to get our attention. It turned out, of course, they were prostitutes. Neither Harry nor I, let's say, were gullible enough, because we were warned that this sort of thing happens. But there was one of the participants who got entrapped this way. I think the reason we were told to be wary of these encounters was because these women were officers. Because we were from Lincoln, we were warned to be careful and not give out any information. So, Harry and I just shrugged it off and walked away.

CHAPTER 8

Francis Bitter National Magnet Laboratory, 1958–1981

Proposals for the Magnet Lab were written in 1958

That summer of 1958, before I had left for my six-week grand tour, we started planning for the Magnet Lab. In fact, you [*Don Stevenson, interviewer*] and Henry Kolm [7.37] were asked to work with me, and we had written a preliminary proposal. But because I was absent, you and Henry were asked by George Valley [8.1] to go down to Washington, DC to make the presentation. Apparently it didn't succeed too well.

So when I got back, George Valley collared me - I think when he came back for a visitation to MIT - and said to me, "Look, this is going to have to be done right, and you are going to have to make the presentation." So he and I agreed that I would line up John Slater [5.5], Harvey Brooks [7.20], Francis Bitter [7.38], and I think the other one who would come with us was Henry Fitzpatrick [8.2]. So when I came back, you and I and Henry Kolm wrote the formal proposal, which we submitted to the Air Force. This was a much better document.

This proposal contained a lot of my dreams for the Magnet Lab. Not only about building a bigger, better magnet laboratory, but scientific content of some of the things that I had envisioned we would do. We could do much better work on the magneto-optical effects, on cyclotron resonance, on the study of magnetic materials, particularly Foner's [7.13] work, and also the ferrite work that was going on at Lincoln Laboratory with Tannenwald and Artman [7.5] that I had started.

Proposed work for the new Magnet Lab

I anticipated the possibility in this proposal that the Magnet Lab would also engage in plasma physics, with the possibility of contributing to nuclear fusion studies.

I already, in 1958, visualized an ideal plasma machine. In fact, to be honest, we had built a toroid at the Air Force Cambridge Research Laboratories with Fundingsland [6.1], and I was thinking about high field magnets for plasmas ... long solenoids ... and I was thinking at the time, not of the toroid, but essentially a toroidal mirror machine, something that was later built by Oak Ridge. But that's the one I had in mind. At that time, I had never heard of a tokamak [8.3], so I was thinking of a toroidal machine. And I was actively thinking of cyclotron resonance heating, because I had thought of cyclotron resonance heating in my days at AFCRL.

Just before I left AFCRL, I had built a magnet, got high-power magnetrons, and I was planning to do cyclotron resonance heating of plasmas in a magnetic field. But I never got to do it because I decided to leave. So, I was already dreaming of cyclotron heating of electrons and plasmas in a magnetic field. This dates back to 1950 at AFCRL. So, in fact, we tried such an experiment at Lincoln, with Rosenblum and Meyer [7.15], for ion cyclotron resonance. When I discovered we were too late – Stix [8.4] at Princeton had already done it – we quit.

But the idea of cyclotron resonance heating was something I was planning, and this was one of the things I was thinking about doing at the Magnet Lab, which ultimately we did do, but not quite the way I planned it. But we did do such experiments later on.

So the Magnet Lab proposal anticipated a lot of things. The proposal didn't anticipate some things that came later, although I think in an appendix I said that we'll do things we haven't dreamed of. It was a very well-written visionary plan.

I think on the magnet side, both Francis Bitter [7.38] and Henry Kolm [7.37] helped. So we knew what kind of magnet we would use. In terms of the plant, we had a good plan. We knew what we wanted to build. We did propose, in that proposal, to build a magnet laboratory that would achieve dc magnetic fields of 250,000 gauss.

FIGURE 8.1   Prof. Francis Bitter (left) discussing features of a Bitter-design water-cooled solenoid electromagnet with Dr. Benjamin Lax (right). This photograph was probably taken in Prof. Bitter's magnet laboratory in the basement of Building 4 on the MIT campus, *circa* 1957. (Courtesy MIT Museum.)

That was our goal, and we thought that would extend our capabilities for studies of solids by an order of magnitude in high fields, in the magnetic resonance in magnetic material and cyclotron resonance and the magneto-optical effects. We also talked about transport experiments like the Shubnikov–de Haas effect [8.5]. We thought this would be a great advance. Although we already had a pulsed magnetic field, we recognized its limitations. You couldn't do things in pulsed fields that you could do in dc magnetic fields. That turned out to be a correct prediction. It panned out very well.

Presentations in Washington, DC in 1958

So we became very busy in planning. We went down to Washington subsequently with the people I mentioned. We made a presentation to General Holzmann [8.6]. Holzmann was a very knowledgeable person. I think he had two PhDs courtesy of the Air Force, I think one in physics and one in chemistry or something like that. He certainly was a well-educated man, if not at the working knowledge level, then certainly he was well aware of what's going on in science and technology. I would say he was probably one of the best informed technical men in uniform.

He recognized the capability and the promise of the Magnet Lab, and what it could provide for basic science that would support some of the objectives of the Air Force. At the end of that presentation he said, "You sold me."

That was the beginning. So, in 1958 we thought we were going to get the Magnet Laboratory. In the meantime, with the permission of Carl Overhage [8.7] and Radford [4.6], we were able to use some of the Lincoln money to build some Bitter magnets.

So we decided we'd spend some money on magnets, a couple of magnets, using Lincoln funds, and put them in Francis Bitter's old laboratory in the basement of Building 4 on the MIT campus. That's the time we hired Bruce Montgomery [8.8]. [*He designed and built the first series of Bitter magnets that were ordered in 1958 from Arthur D. Little, a broadly based scientific and engineering company in Cambridge, Massachusetts.*]

We started working seriously and planning for the Magnet Laboratory, thinking we'd get it eventually. In the meantime, in the background, people like George Valley were really campaigning for us, working for us in the back scenes.

Dreams about a new magnet laboratory began in 1955–1956

Sputnik went up on October 4, 1957. We had been talking about the possibility of a magnet laboratory long before that, because we had gotten together with Bitter in 1955–1956 and started dreaming about it.

There was a committee. I became a member of the Seitz committee [8.9] and I started talking about that. It was at that time that some of the Washington people, like Max Swerdlow [8.10] and another fellow at ONR, Wayne ... somebody, we'll get the name [*this was most likely Wayne R. Gruner (1921–2009) who was with the Office of Naval Research at that time*]... right after Sputnik, said to me, "This is the time to make your pitch for the Magnet Lab." That's when we decided to write the formal proposal. Everybody in the solid state community knew that I was dreaming about higher magnetic fields for doing the kind of experiments we were doing at Lincoln.

So between 1958 and 1960, there was a campaign going on, both by ourselves and by our friends in Washington. The Air Force decided, due to General Holzmann, to step forward as the agency to administer the lab. The Navy didn't want to do it because they already had a magnet lab at NRL. The Department of Energy had no interest in it, although they were a member of this Seitz committee. They knew about it.

The people who encouraged me most were Swerdlow [8.10] and company, and I give him a lot of credit. He, I think, was one of the most intelligent and well-informed administrators in Washington, with a good perspective of what was important, and what was important to the Air Force, as well as to science in general. There were a number of these people in Washington who played an important role in allocating money, recognizing whom to support, and giving the money to the right people. They were good administrators and Swerdlow, I think, was one of the best.

Swerdlow was one of our encouragers, and so it was his office at the Air Force Office of Scientific Research (AFOSR) that would take the lead in proposing the Magnet Lab to the DoD. And it went up to the Secretary of the Air Force. It went up as high as him. He approved it, and in 1960 we got the money.

During that time, in the interim between 1958 and 1960, thinking it was a sure bet, both Carl Overhage, who was a very good leader and administrator, and I visualized the Magnet Lab as part of Lincoln. We were thinking that we were writing a proposal for Lincoln Laboratory. But the Washington crowd wanted it to be a national laboratory, and they felt if it were part of Lincoln it couldn't be that.

Now this was very disappointing to Carl Overhage and myself. We wanted it to be part of Lincoln. First, the Solid State Group at Lincoln provided a basis of support in materials. You know it would have made the research much stronger, as we found out later when we became separate.

National Magnet Laboratory to be on the MIT campus

But the politics won out, and eventually it was decided that it's going to be on the MIT campus. The rationale given was that it would be near to all the universities, it would be nearer the airport for visitors. All the reasons that made sense. And to some extent, when we operated, it worked the way everybody hoped it would. So this was the idea.

So we realized it was going to be on the campus, so we decided to shift our activities to the campus. We had teamed up with Francis Bitter in the study all along. We had already spent some Lincoln money to build magnets for his laboratory on the campus. And we were building more magnets in anticipation of the new Magnet Laboratory.

In 1960, we got the contract. In the meantime, I was very busy writing the book on ferrites with Ken Button, working very hard with Zwerdling and Laura Roth on magneto-optics, and expanding the studies on that.

New Magnet Lab starts out in the Armory building on Massachusetts Avenue

In 1960 we began operations. We were officially called the National Magnet Laboratory, and we set up headquarters in the Armory building that was across the street from MIT [*Building W31 on Massachusetts Avenue at Vassar Street*].

Don Stevenson was assigned to be full time down there as he had spent most of the time in planning and working with Jackson and Moreland [*an engineering firm in Boston*] in planning the thing. So now the plans were to be more serious.

We appointed a committee [8.11], of which we made Francis Bitter chairman, although I had the final say because I was appointed Director. In fact, George Valley [8.1] asked me if Bitter should be the director. I said, "No." I said, "This is my baby. This is what I want to do ultimately."

Siting of Magnet Lab in the five-story Ward Baking building on Albany Street

We formed a committee [8.11] and we started planning with Jackson and Moreland. And since it was going to be at MIT, we began to look into where to put this laboratory. We finally decided the best place would be on Albany Street if we could buy the Ward Baking Company building.

Stratton [8.12]. I'm going to give him credit. That's my recollection. Stratton was really gung-ho for the Magnet Lab, and for it to be part of MIT, and so he was instrumental in buying the Ward building.

We went to the architects at MIT, and they wanted to tear down the Ward Baking Company building and build a spanking new building. But

FIGURE 8.2   Organizing committee for the new National Magnet Laboratory. Bottom row, L-to-R: Prof. Francis Bitter, Dr. Benjamin Lax, and Dr. Donald T. Stevenson. Top row, L-to-R: Dr. Henry H. Kolm, D. Bruce Montgomery, and James M. West. Lax is the Director of the new laboratory. Bitter is responsible for the design and construction of the new laboratory, and is head of the scientific advisory board. Stevenson, formerly Leader of the Semiconductor Physics Group at Lincoln Laboratory, is Assistant Director. Kolm is intimately connected with the original design concepts for the new laboratory and its magnets. Montgomery will design and develop water-cooled magnets and, with Bitter, will design the high-field 250,000 gauss magnet for the new laboratory. West is Assistant Director for Administration. (Reproduced with permission from a 1960 brochure on the MIT National Magnet Laboratory.)

our engineer, Crawford Adams of Jackson and Moreland, said that was an unwise decision. It would be much better to keep the old building and renovate it. You'd have much more space. The only new building you would put in would be the housing for the generators. We went along with it.

Choice of generators

Then we began to plan the generators. Henry Kolm wanted to build a homo-polar generator. We studied the problem. We went both to Westinghouse and GE, and we were given proposals by both of them for more conventional generators.

In the meantime, Francis Bitter and I went to visit Princeton where they were building a fusion machine. They were building a similar generator, a little bigger, with a flywheel for plasma research for fusion. And since I had, in the proposal, written in fusion, I decided that this was a good feature to put into the motor-generator set so we could have a pulse system for fusion.

This was apparently a wise decision and Francis Bitter concurred. He thought it was the right thing to do. So the committee met and mostly, I think, the plans and recommendations were made by Crawford Adams, and we accepted most of them.

We rejected the homopolar generator. We didn't want to do research on power plants.

The decision between Westinghouse and GE was up for grabs. We didn't know which one. But it turned out GE had committed a very serious offense against the government. It overcharged. [*Well, all the electrical manufacturers were nailed on that one.*] But GE in particular. So they offered us a deal to buy this generator set for about a million, which we thought was a bargain. I think Westinghouse was higher. Well, we preferred the GE system anyway. So we bought it and we got it for a song, and this gave us a lot more money for the rest of the system. And we began to do a lot of engineering.

We dug up the MIT athletic field and put in two 4-foot-diameter pipes for the cooling system across into the Charles River. We built a copper bus bar system and the whole power plant.

Experiments begin

In the meantime, we started doing research immediately across the street. We built more magnets, which we bought from A. D. Little, built by Bruce Montgomery [8.8]. We had been installing magnets earlier, but now began installing more and, using some of the money, we renovated Bitter's laboratory, and now we had our own magnets at 100 kilogauss to do various experiments.

Some of the experiments that were done were in spectroscopy. One of the experiments was the thesis by Yaacov Shapira [T1] on the giant quantum oscillations in gallium.

We were doing Faraday rotation in those high field magnets, and we even had a cyclotron resonance experiment (which I'll talk about some more) on diamond, which was in itself very interesting. We had started it at Lincoln and decided we needed higher fields, so we used millimeters and it was the first successful experiment in diamond, p-type diamond [8.13].

We also began to do magnetic experiments. We transferred from Lincoln Henry Kolm, Si Foner, and Ken Button, and they became members of the Magnet Lab. So we were into magnetism. Si Foner began building his magnetometers and we were into semiconductor research, optical research.

We had postdocs: Nishina [8.14] who did the Faraday rotation on semiconductors, Runciman [8.15] who did optical studies of paramagnetic rare-earth impurities in host crystals, and we had two others, Carl Stager [8.16], and Carl Pidgeon [8.17] from England.

I also had a couple of Polish postdocs who began working with me on the theory of Faraday rotation, the semiclassical theory, which later proved to be quite correct, although our first publication had a mistake in it. Nevertheless, we discovered and observed the phenomenon. One of the students was Jerzy Kolodziejczak, who did the Faraday rotation theory with me. [*The other was Wlodzimierz "Wlodek" Zawadzki [8.18], also a theorist.*] Sosnovsky [8.19], who was the head of the semiconductor institute in Warsaw, sent me these two young postdocs.

These were very exciting times. We were engaged in so many things besides the Magnet Lab.

Magnetoreflection experiments

In the meantime, Nishina [8.14] and I – mostly Nishina – did the reflection experiment of interband Faraday rotation in various semiconductors, including a higher transition in indium antimonide, which we published as a paper [P8.1] in the proceedings of a conference [*Conference on Semiconducting Compounds, June 14–16, 1961*] held at the General Electric Schenectady Research Laboratory. The proceedings appeared as a special issue of the *Journal of Applied Physics*. This had many interesting papers on III-V semiconductors.

I presented an invited paper on magnetoreflection [P8.1], which included the work of George Wright [8.20, P9.1, P9.2, T55] and myself, on the interband transitions in indium antimonide. This actually preceded the work of Pigeon and Brown [8.21], but it demonstrated the feasibility of the magnetoreflection technique. This was the first observation of the magnetoreflection.

Alvarez incident

There's a very interesting incident that occurred just before we were ready to build the Magnet Lab after we got the award.

Alvarez [8.22] came to MIT and talked to the vice president. It was General McCormack [8.23] who gave me a call and said Alvarez, who of course was a big shot in physics, had reservations about building the Magnet Laboratory. He thought that there was a discovery at Lawrence Livermore that would make it unnecessary.

So he came and we had a luncheon. He told me that Dick Post [8.24] at Lawrence Livermore had built a liquid-nitrogen temperature aluminum magnet that outperformed copper and would make a high field magnet possible without water cooling, without doing what we were doing, and would require less power.

He told me how it behaved, and I told him I don't think you know what … I didn't tell him in so many words, but I intimated he didn't know what he was talking about, because he didn't know how copper and aluminum behave with temperature. This is something we had studied. In fact, Henry Kolm and I had looked into this and I didn't think it was true.

It was in the spring of, I think, 1961. We were already in the process of planning it. And I told him I'm going to be at the Washington meeting and I'm going to meet with Dick Post [8.24]. I went and talked to Dick Post, and Dick Post confirmed my analysis and said it was a lot of nonsense. He didn't think it would do. So we forgot about it, and we proceeded with our work without worrying about it.

But it's interesting how people in one field who are experts aren't experts in another field. And I've seen this happen time and again, as we'll discuss later … some other things that took place eventually in the laser field.

Cyclotron resonance in diamond

One of the most important results was the cyclotron resonance in diamond in which we observed two peaks. And then we did the analysis, which turned out to be incorrect, but at least it qualitatively predicted that there would be another resonance at higher field, and we used the Bitter laboratory, the old Bitter laboratory, to observe it. This was reported by Rauch in 1961 [8.13].

This was the beginning of high field experiments in semiconductors. Later on we did semimetals, but that will come in a later story.

New National Magnet Laboratory, MIT Building NW14 on Albany Street, opens in 1963

In 1963, the new Magnet Laboratory was finished. We had a big opening, and we began working with the larger generators.

One of the big accomplishments, which was reported at that earlier meeting [*1961 High Magnetic Fields Conference at MIT*] was the design of the 250 kilogauss magnet [7.68]. We achieved the 250 kilogauss magnetic field. We cheated a little by putting in a pole piece inside, which then achieved the 250 kilogauss field that we predicted. Without the pole piece it was closer to 225 kilogauss. The extra 25 kilogauss we got from the permandur pole pieces we put in inside the magnet, which is a high magnetization material, and that boosted the field to a quarter of a million gauss. [*Permandur is a cobalt-iron soft magnetic alloy with equal parts of cobalt and iron.*] And of course you could do certain type of experiments in those.

But the other feature of the Magnet Lab was, not only did we make high field magnets, but we made magnets with different configurations for magneto-optical studies, like a two-coil magnet that can give you either the Faraday or Voigt geometry for optical studies. These were, of course, very useful magnets for all variety of experiments that people did.

First graduate students to work in the Magnet Lab: John Halpern and Yaacov Shapira

We opened the door to visitors. At the same time, we also began supporting a number of MIT physics graduate students, some of whom worked with me and some with my associates such as Foner. Among the first students who worked there were John Halpern [T2] and Yaacov Shapira [T1], both of whom did some very interesting work.

Yaacov Shapira did a very elegant work on the semimetal gallium, in which he observed the quantum oscillations that were predicted by a couple of Russian theorists. And it demonstrated the phenomena that they had envisioned theoretically [P8.2].

John Halpern did interband Faraday rotation in germanium and silicon [P8.3], which was essentially an extension of the work we started with Varian magnets at Lincoln Laboratory with my student Dick Brown [8.25]. Brown was the first to observe [*the anomalous change of sign of*] Faraday rotation [*at wavelengths near the energy gap*], which later I interpreted as interband.

So we were now doing interband Faraday rotation in indirect materials, which turns out to be much better than the magneto-absorption measurements that Zwerdling first observed. We could see the structure better because the Faraday rotation essentially is almost the derivative of the absorption. It's a dispersion phenomenon rather than an absorption phenomenon. And so you could observe it below the energy gap, but you could also observe it above the energy gap. This was a very interesting development.

These were a couple of the first experiments at the Magnet Lab to be done by students, Shapira and Halpern. I think some of them we did in the Bitter lab, and later we did some in the new Magnet Lab after it opened.

Improved magneto-optics in indium antimonide

It was about that time that Carl Pigeon [8.17] joined us. He and Dick Brown [8.25] teamed up to do a much more elegant job on the interband magneto-optical effect in indium antimonide than we did in the late 1950s, and they correlated their data with Evan Kane's theory [8.26]. It turned out to be a seminal paper [8.21].

They did a very low temperature, high field experiment, similar to what Zwerdling did, except with much higher fields and higher resolution. They found all the parameters, both of the holes and the electrons, and so this paper is referred to a great deal. But that was one of the things that we predicted would be one of the benefits of using higher fields. We would get structure in greater detail.

The theory fit the experiments beautifully. They used a complete 8×8 **k·p** theory in a magnetic field to interpret the data. This was similar to the work that Laura Roth did on germanium, except this turned out to be much more precise, with a wonderful fit to the indium antimonide data, which demonstrated the non-parabolicity and determined the mass parameters of both electrons and holes.

So we began to understand much better the band structure of these semiconductors, indium antimonide and others. Other III-V semiconductors are very similar to indium antimonide, particularly the valence band and the conduction band when they are direct transition semiconductors. So the cyclotron resonance and the magneto-optical effects were instrumental in opening our understanding quantitatively of the band structure of these important semiconductors.

This was one of the intentions of the Magnet Lab, to make this kind of a contribution. And this was one of the early big successes of the high magnetic fields.

Superconductors in pulsed magnetic fields

The other thing that occurred, just about the time of the conference [*1961 High Magnetic Fields Conference at MIT*], was that the Bell people developed these high field superconductors [7.68]. And one of the Magnet Lab programs was to measure the critical fields in these superconductors. Our dc magnets were just in the right range for niobium-tin and niobium-titanium, on the order of 100 kilogauss plus, between 100 to 200 kilogauss.

So, many visitors came and measured these superconductors, as well as Si Foner, in his magnets, his pulsed magnets. At Lincoln we built a pulsed magnet system, which started us in this field, using a 10,000-joule condenser bank that came with the power supply that I got from Raytheon.

But then, when we went to the new Magnet Lab, one of the magnet projects was for Si Foner to build a 100,000-joule, larger condenser bank, a larger pulsed magnetic system, which he did. He used this very usefully for studying critical fields in superconductivity and other magnetic phenomena at high fields. We also developed a lot of instrumentation to go with the high field magnets, magnetic measurements.

An important development that happened was, when the high field superconductors were discovered, about the time of the meeting, a number of people, including myself, certainly Bruce Montgomery, began thinking about building what we called the hybrid magnet. [*The hybrid magnet is a conventional water-cooled magnet surrounded by a high-critical field superconducting magnet.*] The concept of the hybrid magnet must have occurred at conversations a number of us had amongst ourselves, but certainly it occurred in the Magnet Lab, and Bruce Montgomery of course was the one who would build it.

We recognized that the highest critical field achieved with a superconductor is about 170 kilogauss with niobium-tin. Of course, in a large magnet you never get that. But if the magnetic superconductor is outside, the fringing fields of the water-cooled magnet won't quench the magnetic field of such a large volume magnet.

We began planning on that, and predicted that, in addition to the water-cooled magnet which could achieve in excess of 200 kilogauss in a usable volume, we could add 50 to 100 kilogauss to that and exceed the 300 kilogauss limit. This was the goal. This was a major project which, in some ways I now regret, because the lab now placed more emphasis on the magnet development at the expense of some of the research.

Magneto-optics of bismuth and helicons at the new Magnet Lab

In addition to the work on magneto-optical effects in semiconductors, which was one of the things we had anticipated with the discovery of the magneto-optical effect at Lincoln Laboratory on bismuth, Brown, Mavroides, and I decided to continue this work in the Bitter magnets. Indeed, I think we did some of this at the old Bitter magnet laboratory and produced spectacular results, from which we got mass values of electrons and holes in bismuth.

Another interesting phenomenon that was investigated, that had been predicted by Aigrain [7.56], is a plasma phenomenon that's well known. He called it a helicon, a helicon effect. This is an interference effect in a finite geometry, and this was observed by Furdyna [8.27] at these high fields, up to 100 kilogauss, in a most spectacular way. They too give you information about the free carriers in semiconductors like indium antimonide. They essentially explore the dispersion effects in a plasma, in a cold plasma in magnetic fields, using microwave techniques and in a semiconductor slab.

These were some of the early successes at the Magnet Lab shortly after we opened the laboratory in the period of, I think, 1964–1965.

Bitter magnets and pulsed magnets for scientific research

At the Magnet Lab we proposed a whole host of experiments. We proposed both magnet technology and basic experiments. We would advance the Bitter magnet technology, pulsed magnet technology, and do experiments with both of them.

Foner was the one who took the responsibility for the pulsed system. We built a bigger pulse system for him at the Magnet Lab than at Lincoln Laboratory. In 1963, when the Magnet Lab opened, we already had Bitter magnets to go into the cells.

We began a whole series of experiments. Magneto-optics and spectros-copy in magnetic fields was one of the important areas. Runciman [8.15] specialized in spectroscopy of rare-earth materials. We, on the other hand, began magneto-optical effects in semiconductors which we discovered at Lincoln Laboratory. There were some biological experiments by visitors.

We had now eight small cells and two big cells. We had many magnets. Gradually different experimental setups were built, some which were there for quite a while, others which could be taken in and out [8.28].

Aggarwal joins Magnet Lab in 1965, sets up infrared modulation spectroscopy

In 1965, when I became a professor, I decided that one of the things I wanted was a spectroscopic setup. Roshan [*Roshi*] Aggarwal, who came from Purdue, joined me that year, and that was one of the first things we did.

Roshi's professor at Purdue was Ramdas. Roshi did infrared spectroscopy for his PhD thesis. He started doing spectroscopy using modulation techniques. He redid some of the things Zwerdling did, but at higher fields and with this improvement in the spectroscopic technique.

He built an infrared spectrometer with the usual Perkin-Elmer spectrometer, a grating spectrometer, with low temperature dewars, and, for the first time, he built a little window with a germanium attached to it on a heat sink, and we redid the Zwerdling experiment of the indirect transition [P8.4, P8.5]. But this time we had the germanium sample mounted on a piezoelectric transducer, so we could modulate the transmission, and in this way we got the derivative of the transmission spectrum, whereas the Zwerdling data was a step function which was hard to resolve.

The indirect transition in germanium is a step function, a series of step functions and it's broadened. It looks like a series of steps that have been broken because the slope of the step isn't a right angle but an "S" shape. But when you take the derivative they show up as peaks. So we were able to go to much higher energies. I developed a **k·p** theory and, as I said, it checked pretty close but not quite close enough because I neglected the Coulomb interaction, which plays an important role in the interband transition.

Miura [8.29] in Japan did cyclotron resonance at high fields, which also took him high into the conduction band, and consequently he could see the non-parabolicity without the Coulomb effect, and there my theory and his experiments agree pretty well.

This was one of the first projects with Roshi among many which I'll discuss. In fact, he and I collaborated on a lot of very interesting and successful programs in the spectroscopy of semiconductors.

Mossbauer effect

Another experiment that we started was the Mossbauer effect. This was a phenomenon discovered a few years earlier. This was a high resolution experiment in a nuclear transition. I thought it would be a nice experiment to do in a magnetic field.

Don Stevenson and I decided to try it in a Bitter magnet, but then we realized that the vibration of the water-cooled magnet made it impossible. Ultimately we agreed that, although this was a good experiment, it should be done in a superconducting magnet. So, in addition to the water-cooled magnets, we planned on acquiring or building superconducting magnets for such experiments that required very low noise or quiet fields. The water-cooled magnets with the water pumping through could not provide that.

My objective was to demonstrate that the photon had a spin, or a rotating photon had angular momentum, because it transferred it to a particle. But we never did that experiment.

We hired Norman Blum [8.30]. Norman Blum ultimately set this up, and this became another major project in the laboratory. This was later taken up by Dick Frankel [8.31], and was a very productive tool for the study of magnetism. They exploited the Mossbauer properties of iron and other magnetic materials. Frankel and company with Foner were studying antiferromagnets and other magnetic systems using the Mossbauer effect. Frankel also got into looking at the Mossbauer effect of biological molecules, which contained iron and various other materials, which I think was very interesting work.

Effect of a magnetic field on a semiconductor laser diode

Another project that we decided to study was to put semiconductor lasers in a magnetic field. In 1957–1958, we discovered [P8.6] that the absorption, particularly in low gap semiconductors, say, indium antimonide, was very dramatically affected by a magnetic field, and this was why I wanted indium antimonide to be the first laser.

We put the laser, a diode laser, indium antimonide, in a magnetic field [P7.26]. We did discover, just as I had predicted, that the threshold would be reduced with the advent of the magnetic field due to the apparent singularity in the density of states, which I believe wasn't real. No singularity exists in nature.

Sometime in the 1960s I had a student, Barry Sacks [T3], who was an electrical engineer. I told him that we ought to try to explain this phenomenon, so we worked on the broadening of the density of states. We developed a phenomenological theory, which we found was equivalent to a more sophisticated theory developed by a Japanese theorist, Kubo, the Kubo density of states. So we had arrived at it in a phenomenological way, and sure enough it did explain the behavior of the magneto-optical laser, as we called it. This was a good idea.

Other people brought in lead-salt lasers and others, and observed various phenomena in a magnetic field.

Lasers were studied in high magnetic fields by the Lincoln group [8.32], and this is what I meant, that the combination of Lincoln and the Magnet Lab would have proved to be very good. One of the very interesting experiments was done by Butler and coworkers on looking at the magnetic field effect on lead salt lasers, in which they saw a very interesting effect. The emission peak shifted considerably. They went up to 100 kilogauss and they even saw the spin splitting. They measured the g-factor for the first time and with great accuracy in these materials [8.33].

Superconductors characterized at the Magnet Lab

The other big effort was the measurement of high field superconductors, particularly with the dc magnets. A number of visitors did that. These were some of the bread and butter programs. Si Foner did some of this with pulsed magnets.

So, we had a wide range of measurements with high magnetic fields for a large variety of materials. Once again we made important contributions to an area which ultimately became a practical technology.

Nuclear Magnetic Resonance

Another area of high field research that I thought would be appropriate for the lab would be nuclear magnetic resonance (NMR).

This idea I got from one of my visits to Bell Telephone Labs. A man named Shulman [8.34], who was a well-known nuclear resonance spectroscopist, in the mid-1960s was working with superconducting magnets. Most of these were niobium-titanium magnets that only went to about 50 kilogauss, and at 50 kilogauss you could resolve certain lines, but other lines were unresolvable.

I said to him, "How would you like to do NMR at 100 kilogauss or higher?" He said he'd love to. And I said, maybe we can do that in a Bitter magnet. And he told me that, in NMR, the resolution goes, I think, as the cube of the frequency, so that you would get a big gain in going to twice the field, in fact a little more than twice the field, you could get an order of magnitude, and this was important for many organic systems and biological systems. We recognized this possibility. I said this looks like an exciting area.

So I came back and talked to two people. One of them was Bruce Montgomery and the other one was John Waugh (1929–2014), a professor in the MIT Chemistry Department whose expertise was in NMR. I suggested maybe they collaborate and begin such a thing. It never panned out.

We explored the idea and came to the conclusion that this would be a good major project in the laboratory. Ultimately, we recognized that a superconducting magnet would be better than the dc magnet because, again, of acoustic noise. Because the lines were narrow, the acoustic noise would bother us.

Then Neuringer [8.35] got very interested in this possibility and decided he would make it his career. Indeed, he began working on it, built up a program, and came up with the idea that he would build a niobium-titanium high-field NMR magnet using pumping on it that would go above 100 kilogauss. A similar project was also being contemplated at Oxford where they would use niobium-tin, which we thought was a more difficult technology but had a higher critical field. This was done by Oxford Instruments.

So that became another major project in the laboratory. That came out of a visit. It became a very successful program, and became one of the important facilities at the Magnet Lab in the 1970s. The NMR work was supported by the National Science Foundation. In 1972 we hired Bob Griffin [8.36], and he and Neuringer began this work. Bob Griffin was a postdoctoral student of John Waugh's, and so he was the ideal man for this.

We started the high field superconducting magnets for NMR. Neuringer set out to build the first hundred-kilogauss high-frequency NMR magnet, niobium-titanium. I think he made very good engineering decisions, and I think it was the first one. The competition was niobium-tin, which had many more difficulties. By pumping on the niobium-titanium, they were able to get to 100 kilogauss. I think this was the first such high-frequency NMR magnet. It became part of the facility that he ultimately built up.

CO_2 lasers enable new experiments

We also began projects on plasma physics. The CO_2 laser, the transversely excited CO_2 laser, had been invented. Charlie Chase [8.37], Meservey [8.38], and Maxwell [8.39] joined the Magnet Laboratory. I asked Charlie

Chase to build us a transversely excited CO_2 laser. I wanted to use it for plasma physics and some solid state physics experiments.

Indeed, we built CO_2 lasers, both Q-switched, and these pulsed high-power lasers. Ken Button began to study multiphoton absorption in semiconductors using the Q-switched CO_2 laser.

We began some nonlinear optics with the transversely excited CO_2 laser. Soon thereafter it became apparent that it was not a wise thing for us to build our own lasers, because there was a company in Canada, Lumonics, which was building these lasers with a grating. This I did not know until I visited them.

Far-infrared spectroscopy with FTIR spectrometer and cyanide laser

In the 1960s we also had a visit by Gebbie [8.40], and he brought two instruments that were well-tailored to doing spectroscopy experiments in high magnetic fields.

One was a Fourier transform spectrometer in the far infrared, which covered the wavelength range from about 100 microns to a millimeter. And we began doing experiments on magneto-plasma effects, Zeeman effects on polarons. The other thing that he brought in with him was the cyanide laser, which operated at around 300 microns. This was a discharge laser. So, for the first time we had a far-infrared laser, a coherent source, with the high magnetic fields, and we began to do cyclotron resonance experiments.

One of the first experiments that we did with Ken Button was to do cyclotron resonance using Gebbie's cyanide electrically excited laser at 337 microns.

The pulsed laser for cyclotron resonance was then picked up both by the English and the Japanese as a standard tool, although I think Button, Gebbie, and I pioneered that. And that began to replace the microwave sources. It gave you a tool for measuring a larger variety of semiconductors.

We did some very interesting experiments. The Fourier transform spectrometer was used primarily later by the students. For example, we used it for such studies, I think, with Dick Stimets [T5], on the magneto-plasma effect. We studied the Zeeman effect in cadmium telluride, which was the thesis of Dan Cohn [T11]. This was done with the Fourier transform spectrometer, and we saw the polaron effect. Dave Larsen [8.41] did the theory for this because this was his specialty. And this was a nice marriage.

One of the big successes was the polaron experiment suggested by Larsen at Lincoln, which my student, Waldman, did for his thesis [T6], and which we ultimately repeated, and which I explained theoretically as the best measure of the electron-phonon coupling.

Magnetic separation project

Other applied experiments were done by Henry Kolm. For example, he began a project on magnetic separation, which was a highly applied program, a very worthwhile one, which ultimately was adopted by industry. It was a unique idea. This too became quite successful.

Kolm and Kelland worked on it and demonstrated the feasibility. They would take dirty clay, take out the magnetic impurities, and it became white. This also, I think, became a commercial item. So the Magnet Lab began this area.

To some extent, some of this applied work was not viewed by the Institute quite in its proper light, although today the climate is different. They felt the Magnet Lab should be a basic research facility, but I felt that applications were just as important as the basic research we did.

Low temperature experiments: adiabatic demagnetization, magneto-tunneling in superconductors

The other kind of experiments we were doing were low temperature experiments, like those on adiabatic demagnetization done by Maxwell. This program was ultimately discontinued while we were still with the Air Force because of the Mansfield Amendment. That was discontinued, unfortunately, but I thought it was an appropriate program.

But the most interesting program that came out of the Meservey group [8.38] was the tunneling experiments in superconductors in magnetic fields, spin conserved tunneling. This became also another successful project. And that's still going on. [*Yes, their postdoc, Jagadeesh Moodera, is now one of the top people in that field.*]

But spin polarized tunneling is, I think, something that Meservey discovered, and is an important tool for the study of superconductors and Josephson junctions and other kinds of superconducting tunneling.

This was discovered by Bob Meservey and studied, with Paul Tedrow, extensively. This was a very nice piece of work. They even made their own films in the laboratory. This, I thought, was one of the unique and important areas of science.

Management philosophy disagreement: magnet technology *versus* scientific research

My roles at Lincoln and the Magnet Lab during this period were intertwined. Until 1965, I was directing both the Solid State Division at Lincoln and the Magnet Lab. So we had a good relation.

In fact, that was my intention for creating the Magnet Lab, to be part of the solid state effort at Lincoln, because I felt just having a lab that just is a facility is not as good as if it were part of a larger effort where you provided material, scientists who could take advantage of the Magnet Lab.

And it turned out that this was a correct assessment, but unfortunately the politics demanded that the Magnet Lab be separate from Lincoln. It was a great disappointment for both Carl Overhage and myself. And I saw the weakness in this, in that we became a national facility rather than a laboratory.

Of course, some people like Al Hill [5.3] and others accused me from time to time that it's Ben Lax's laboratory, which of course was not true. Not everybody did the experiments that I suggested or participated in. In fact, there were a lot of independent efforts, and, as the Magnet Lab became more mature, this became more apparent.

But of, course, the emphasis on the Magnet Lab as a facility was from the sponsors, AFOSR, the outside committees, which I always viewed with great reservation. My feeling is, no committee that comes in for one day or two and gives opinions off the cuff can substitute for a well-organized laboratory run by people who live with the program and the work throughout the year. I believe the Magnet Lab personnel, including myself, were smart enough and were, by and large, in a much better position to decide what was the best thing for the laboratory.

Of course, we didn't always prevail. For example, we conceived of the hybrid magnet and we started that program. The committee said, particularly with the encouragement of the NSF sponsors, that we should emphasize the magnet technology more at the expense of research. Well, I think this was a bad decision.

My position was that the magnet program was a capital outlay, and should be in addition to, not at the expense of research. In other words, to me, adding another 50 kilogauss to a 250 kilogauss magnet at the cost of a million dollars seemed very stupid, and I believe it so today, because it didn't create new programs, new ideas. It merely pushed the highest magnetic field just a little higher. It didn't make any new experiments possible.

This is the kind of thinking that goes on in Washington, and apparently the advisory committee bought it. But I think it was a very poor way of spending money. Nothing substitutes for the creative talent of the postdocs, students, as well as staff, who can do highly original work and come up with new ideas, new instrumentation for using the magnets. But I never won that argument. This was particularly true under NSF, I think.

Research flourishes in the AFOSR period 1960–1970

Under AFOSR for the first ten years, we had more freedom to make our own decisions, although the budget never was raised. NSF took over in 1970, and we did start going out for external money. Up until then, we simply used the money given to us by AFOSR, which amounted to something like two to three million.

But we flourished quite a bit under AFOSR. Under AFOSR I had more students, less emphasis on magnet technology. We had postdocs from all over the world, including Vrehen [8.41a] from Holland, Zawadzki [8.18] from Poland, and Nishina [8.14] from Japan. These people made very important contributions, and each of them had a different program.

We covered a broad range. We kept up the magnet technology at a modest and appropriate level. But the philosophy under the AFOSR management that I tried to establish was that you build magnets not for their own sake, but for new kinds of experiments.

For example, for the magneto-optics programs, we built these double Bitter magnets in which you had optical access both along the direction of the magnetic field and transverse to it, and this made possible certain kinds of spectroscopy as well as other experiments.

Even in the plasma area later on, we used these double Bitter magnets with great advantage. So the magnet technology was the servant to the research, rather than the other way around, which is what prevailed later under the auspices of the National Science Foundation.

The period under AFOSR sponsorship was I think the most enjoyable. This was the period between 1960 and 1970. That's when we acquired most of the students and postdocs, and we did highly original work.

Millie Dresselhaus becomes an important visiting scientist at the Magnet Lab

Visiting scientists included MIT Professor Millie Dresselhaus [7.44], who was participating in magneto-optics, including the Raman studies that we started with students in the 1960s with Roshi Aggarwal. She took over that apparatus and began to do some very interesting work on europium chalcogenides, which later one of my students, Sam Safran [T22], a very bright young man, interpreted theoretically. He's now the Dean of Science at the Weizmann Institute in Israel. This became an interesting field of high field research at the Magnet Lab.

There was earlier work on antiferromagnetic materials. The chalcogenides went through several magnetic phases as you varied the temperature, and the spectra changed, and essentially you had a magnetic lattice which my student, Sam Safran, explained.

Millie Dresselhaus and some of her students began the magnetoreflection studies in graphite, analogous to the work that we started at Lincoln Laboratory with bismuth with Mavroides and Dick Brown. This work became a major project at the Magnet Lab.

Millie Dresselhaus sets up a nursery in the Magnet Lab

One day, while Millie was running an experiment, she called me in to show me something, and in one of the spare rooms there was one of her little babies. Apparently they made a nursery where, in between magnet runs, she would nurse the baby.

I think Don Stevenson made this accommodation, which I thought was very amusing, but also very appropriate for the kind of spirit that prevailed in the Magnet Lab, a very free spirit, a very cooperative spirit.

I think this was what was nice about the operation during the AFOSR period [*1960–1970*]. We had more freedom to do things, and we weren't too heavily pressured to be primarily a national lab, although we fulfilled that role quite well.

Trouble satisfying both NSF and MIT

But at the same time, we also were an MIT lab, which was something that the administration, typically people like Jerry Wiesner, desired. He wanted the Magnet Lab to be like RLE. But this of course was not truly possible because of the limited budget that we had, together with the pressures to be both a facility and an MIT laboratory. These competed with one another. Under AFOSR it was reasonably well balanced, but apparently when NSF took over, we couldn't satisfy both requirements, and the shift went to being a facility, and we were criticized for having too many MIT people using the Magnet Lab, which meant students and members of the MIT community.

That I think was a mistake. If they wanted it to be a national laboratory and an MIT laboratory, we should have fought for more funds. But this did not work out. In fact, when NSF took over we took a half-million dollar cut.

I was at an Academy cocktail party and I met Guy Stever, who was head of the NSF at that time, and I told him, "You're expecting more and more for less and less." It was true. In other words, we couldn't satisfy both communities, and that was unfortunate. I think they both had legitimate desires, but to be able to do that you need a much bigger budget, particularly with the emphasis on magnet technology.

Mansfield Amendment shifts Magnet Lab support from AFOSR to NSF

In 1969, Senator Mansfield made a criticism of the services supporting basic research. He thought this wasn't the proper role, that they should be more applied.

One of the first casualties was the Magnet Lab. In other words, he didn't think the Magnet Lab was an appropriate laboratory for the Air Force to support. It was a basic research facility and therefore it should be supported from somewhere else. The logical candidate, of course, was the National Science Foundation.

Well, NSF decided to take us over, and there was a lot of maneuvering in Washington, I'm told by Jerry Wiesner and others. They all played a role in seeing to it that it was properly transferred.

The Mansfield Amendment also was responsible for transferring from AFOSR the materials science centers at universities to NSF, so both the Magnet Lab and these were put into the same division, and unfortunately it was sometimes viewed they were competing for the same funds. This of course created problems which I can discuss later.

In fact, very often some of the members of our advisory committee were members of the materials committee. There was a committee there within NSF that was represented by all of the materials centers but never represented by the Magnet Laboratory. Yet they were discussing the fate and future of the Magnet Laboratory. It wasn't right. I should have been a member of that group. But that was politics.

Well, NSF ultimately took us over. In the transfer, the budget was reduced by half-a-million dollars, which necessitated cutting out some of the postdocs and others. It turned out to be at the expense of research, and of course later this was aggravated by the shift from research to operation of the facility. But the science continued on.

Seeking outside sources to supplement NSF funds

Under AFOSR we were not encouraged to go outside for funds, but under NSF we were able to go out after other funds. It was difficult to get anything [*additional*] from NSF because they claimed they supported us adequately. But we could get money from other agencies, which we did.

Of course this expanded some of the applied programs as well as the basic program and some programs like magnet technology. We had outside sponsors who wanted us to build magnets for them. We started the program with Neuringer on the superconducting high field NMR magnet, which we started first on the NSF budget. But ultimately this program was shifted to NIH. So it was sort of seed money.

Optically pumped far-infrared lasers in gases

The discovery of the optically pumped laser by the Bell Labs people expanded the number of far-infrared gas lasers. We did a development program on that with Dan Cohn [8.42] and company taking the lead.

This now became a very interesting research topic of its own. We used a pulsed CO_2 laser to excite a number of gases, and the most familiar one was the CH_3F molecule.

This permitted us to get a larger variety of transitions. It could be tuned with a grating. We also studied the physics of this laser, and we used it in breaking down gases, in doing more extensive cyclotron resonance programs, and as a source for plasma diagnostics later when we began a major program at the Magnet Lab.

Research at the Magnet Lab in the 1960s

So I think we're now beginning to discuss the period between 1960 and 1970, the period under the auspices of AFOSR. I think this period was characterized by what I thought was a fair balance of internal research by the staff, students, and postdocs (we managed to hire a few of them), some from overseas, and also by visitors.

But when NSF took over, it was made clear to us that greater emphasis should be placed on visiting scientists and less on the MIT use of the Magnet Laboratory. And obviously, with the funding that we received and greater emphasis on magnet technology, this of course meant that internal research including the number of students, postdocs would be cut in the 1970s.

David Cohen pioneers magnetocardiography (MCG) and magnetoencephalography (MEG) at the Magnet Lab

We had another new program which we started sometime earlier, and this is a very interesting story. In 1969, David Cohen [8.43] came to me from the University of Illinois in Chicago, where he had a magnetically shielded room. Apparently he and some of the people on the faculty didn't get along, and he was looking for a new home. He came and visited me, and he described to me his idea of studying the magnetic field of the human heart in a magnetically shielded room.

I was very intrigued. Apparently he was one of the few people, and maybe the only one that I knew of, who built a shielded room to exclude the earth's magnetic field so he could detect these small magnetic fields that are produced by the heart. I told him I would give him a partial support, and he would set up here at the Magnet Lab. That began this program, and this ultimately became very successful.

He pioneered not only detecting the magnetic field of the heart, but also that of the brain. I think he was the first one to ever do that. And this today is a very important program around the world.

With the SQUID magnetometer in connection with his shielded room, David Cohen's work began flourishing. He did some very nice work, not only studying the magnetic field of the heart and the brain, but he also began some very practical work on magnetic detection of asbestos in the lungs of asbestos miners from Quebec. He demonstrated that this was a feasible tool for detecting the level of asbestos in their lungs because asbestos contained magnetic impurities.

This was one of the important achievements of the Magnet Lab.

Magnetically guided catheters

There was another very interesting program that was started at the Magnet Lab. That was magnetically guided catheters, led by Bruce Montgomery. You put a catheter inside a body, and use an inhomogeneous magnet to push it up the line, to guide it through the body. This I think ultimately became a commercial thing. I don't know how successful. But we did transfer it to industry. That was a very interesting program, which ultimately was supported by the National Institute of Health. Montgomery and

Hale later developed the superconducting magnet for guiding catheters, which was an improvement on the earlier work.

Progress around 1972–1973

Of course we had major activities on magnet research and technology, building new magnets, improving the old ones, working on the hybrid magnets. This was accompanied by a major activity under Larry Rubin, with instrumentation which of course was necessary. He supported a lot of the people who came in, the visitors as well as the people in house.

Laura Roth joined us in 1973, working on magneto-optical theory. We started a new program on cryocables with Peter Graneau. That was a good program. Safran was the student. He joined us in 1974.

We also had a very active program in plasma physics under Cohn and Halverson, with my participating in it. We were working on the interactions of lasers with plasmas as well as on the operation of optically pumped lasers, which we used both as a tool and as a research study. And that's when the molecular biophysics program under Neuringer, who had his NMR magnet, began. In addition to Griffin, he hired Dave Rubin to work on the instrumentation.

So the Magnet Lab had a broad spectrum of activity. This helped to interface with the visitors, who used mostly the facilities and instrumentation that we provided, and who also worked jointly with our staff.

So I thought we fulfilled the goals for the Magnet Lab, although we kept getting constant criticism that we didn't have enough outside participation. This was, of course, not true.

Conference on semiconductor heterojunctions, Budapest, 1970

I'll tell you about the trip to Hungary. It was very interesting. It was in 1970.

It was a conference on semiconductors in Hungary organized by a man named Szigeti. [*The International Conference on the Physics and Chemistry of Semiconductor Heterojunctions and Layered Structures was held in Budapest, Hungary on October 11–17, 1970. G. Szigeti was editor-in-chief of the five-volume proceedings.*] He is a man I met actually at a previous semiconductor conference. He was a fatherly sort of man who had escaped the Holocaust. He was a Jewish physicist, and he was very much taken with me, and was very interested in the work we were doing here at Lincoln. We got to be pretty friendly.

So when he had the conference, he invited me and I decided to go. I was very busy at that time. I think that was the time we were in the transition period from AFOSR to NSF, so I didn't have too much time.

I was doing back-of-the-envelope calculations on graded gap semiconductors and when is it meaningful. In fact, the considerations of being meaningful were very similar to what Esaki was talking about in terms of quantum wells. When the structure in the quantum wells became, let's say, of the order of 100 angstroms, then it began to make a difference in the energy levels. Similarly, when the graded gap, let's say, was graded over something between 1,000 angstroms or less, of the order of 100 angstroms particularly, then it also became very meaningful, because if you make a graded gap material, it automatically has a built-in field because the energy gap changes so that the electrons slide down and the holes go the other way. The electrons slide down the energy bands to the lower gap, and this is the equivalent of a field. It gives you certain advantages in transport devices.

It also is very important for avalanche devices. At that time, I wasn't considering that possibility because I hadn't yet gotten involved with Raytheon in avalanche diodes. But it's a very interesting concept.

I didn't have time to prepare the paper, so I wrote the paper on the airplane and got there and I gave a talk. One of the physicists, I forget his name, was so impressed he says – and I told him how I prepared it – he says, "It doesn't matter, you always give a good and interesting paper." He gave me a compliment.

The night before the conference, Szigeti invited Esaki and me to a night club (it's like a cabaret) at the hotel where we were staying. There they played music. You know, Hungarian musicians are very good with the fiddle. So for Esaki, Szigeti got them to play a Japanese song. For me they played a Hungarian song. The three of us were eating dinner there and having a good time.

Visits Miskolcz and his old neighborhood

The other interesting part of this Hungarian trip was that I decided to visit my home town, Miskolcz. You remember this meeting was in Budapest.

I took a train to Miskolcz. I got off the train and got to where we used to live. The house that I lived in was gone. It must have been blown apart. In fact, there was a street up that lane where I got hit with the stone. That lane became a street now and there were houses on either side.

But the street was there, the church was there, that woodworking shop that used to fascinate me as a child was still there. Then I went to the *Buzater* where the circus used to be held. But now they had closed in buildings where the peasants used to come and sell. Instead of an open area, there were big buildings where they used to drive in their wagons and sell their goods. It no longer was an open field. But I recognized it.

Then I went up to my old school. I went over the main street. Everything looked so much smaller. I mean, then I was a little boy, ten years old, about four feet or less than four feet tall, and now I was five foot five. Everything looked small. It took me no time to recognize things, I knew my way around, remembered every street.

I went over a bridge. There was a little stream just between the main street and the other side where the school was. The school used to be a school for just Jewish boys. This was a secular school. Down the street there was a Hebrew school. We attended both, but the secular school was a nice brick building. Again, it looked much smaller that I remembered, but they had turned it into a police station.

Then I walked further up the main street, and I came to this beautiful park called the *Avas*, which looks like a mountain but it's really a tall hill. It's very pretty and overlooks the city. I just went to see if I could recognize everything.

I went into a restaurant, and I was able to order my food in Hungarian. The language sort of came back to me. I could ask somebody where is this place, where is that place. That was one very interesting side trip.

Gives after-dinner speech in Hungarian

Then at the conference I was at the dinner, at the banquet. I was asked to speak for the foreign guests. So I gave my speech. The night before I had asked Szigeti to give me a Hungarian-English dictionary. I wrote up the speech in Hungarian and spoke in Hungarian to the audience because most of them were Hungarians.

Then one of the English-speaking guests asked me, "What did you say?" I said, although I spoke for about five minutes, I said, "In a few words, thank you." I thanked the hosts for the gracious reception and wonderful conference.

It was a very interesting conference. Esaki gave a nice paper, and I gave a paper in English. Most everybody spoke in English. I wrote two papers on graded gap material, one with Honig [P8.8] and one for this conference. [*An unfinished draft of a manuscript by Ben titled "Quantum Effects* in Graded Gap Materials," which was planned to be a sequel to the 1969 paper [P8.8], is contained in Box 2, Folder 22 of the Benjamin Lax Papers, MC 234, at the MIT Institute Archives and Special Collections, Cambridge, Massachusetts.] I never went back to it, which is strange, because graded gap materials in magnetic fields would have been very interesting.

"Death Rays" and X-ray lasers

In 1972, after the mode-lock laser was discovered, I realized that there was enough energy in a laser pulse to possibly excite enough electrons from the K-shell to the L-shell to create lasing in the soft X-ray region. I worked out the feasibility without suggesting a mechanism for inverting, but in any event, if it was inverted, there was enough energy in the laser to create lasing in the X-ray region.

This paper subsequently was slightly improved by Yariv some years later, but in effect it established the criteria, one of the criteria, for a requirement of X-ray lasers in the soft X-ray. But it didn't identify the mechanism of inversion.

Around that time, in 1976, I gave a paper, an invited paper, at Rome Air Development Center [located at Griffiss Air Force Base, Rome, New York] where they were celebrating their 25th anniversary. I was invited there. And I casually mentioned that an atomic blast … in my oral talk … an atomic blast could produce X-ray lasing. This was triggered by a suggestion by someone from Bell Telephone, who suggested that you could excite, if you had an X-ray source intense enough, you could excite those transitions to, let's say, to the continuum and have lasing.

Now obviously this is fulfilled in an atom bomb because the flash lamp, that is, the flash itself, produces black body radiation that is equivalent to 10 million or so degrees or higher. Obviously it could pump the fragments that come off, and invert the laser, and there could be a traveling or a super-radiant laser produced.

I mentioned this only casually in my talk. But it appeared in an MIT campus publication. Now, this was at the time of the Vietnam War, and the students accused me of working on death rays!

I was mortified! I hadn't even thought about such a thing. But another idea, another possibility that was suggested, but at that time I didn't think of it, which is also true in an atom bomb, was that charge exchange between the fragments could also produce X-ray lasing. This was subsequently suggested by a number of people who were thinking about X-ray lasers.

In 1972, Arthur Guenther [8.44] from the Air Force Weapons Laboratory [*located at Kirtland Air Force Base, Albuquerque, New Mexico*] and I wrote a paper [P8.9] on the feasibility of an X-ray laser. There was a group in Washington that wanted to hear ideas on X-ray lasers. Art Guenther wanted to have such an effort at the Weapons Lab, building on some of the ideas that I had suggested. The Weapons Lab had powerful lasers, and if they didn't, they had the money to buy them.

At any rate, I was scheduled to speak, and Lowell Wood [8.45] showed up. He was the first one called upon. He presented a paper which was almost the duplicate of my paper. Of course he did reference my published paper that I wrote with Guenther. I was very upset about it, and I told the chairman that there was no reason for me to present the paper because Wood presented step-by-step what I wrote in the publication.

I said, "I don't have any more to say except that neither of us identified the mechanism for inversion." But at least we established the criteria, one of the criteria required for an X-ray laser.

We didn't get any support at the Weapons Lab, but Lawrence Livermore did get support, and I decided that was it. I'm not going to work on the X-ray laser. The politics particularly, with Lawrence Livermore with the backing of Teller, was something I didn't want to compete with or participate in, because I thought these people were misrepresenting a lot of things that went on at Livermore.

X-rays from hot plasma in a magnetic field

There was the work of Chase and Halverson, with Dan Cohn, using CO_2 lasers to produce X-rays, not lasers, but X-rays by heating plasmas in a magnetic field. This was very interesting work.

Dan Cohn did some very excellent work in this regard. We showed what would happen if you put a high-power CO_2 laser, a TEA laser, into a plasma that was inside a magnetic field. Now, without the magnetic field the discharge was very diffuse. Once you put on the magnetic field, the discharge became a very intense column of light, about a millimeter in diameter. Dan Cohn did some laser probing and indeed showed that there was a density profile.

There was a density profile in the column, which was least intense in the center and had a hump at the periphery. So essentially it was sort of a double hump. This meant that the plasma formed a dielectric waveguide. One can show this theoretically. Then I derived a magnetohydrodynamic model which explained the formation of this density profile. And that was very nice.

There was another phenomenon connected with that. In addition to forming the profile, the laser essentially bleached the plasma and probably heated it quite a bit. Sunny Yuen and I worked out the theory, which we published [P8.10]. It was essentially the same equation that – at that time I didn't know it – but the same equation that is defined in the soliton.

The soliton is a single pulse of excitation that travels, not like a wave, but travels through it. So it was the same equation and we published it. So it was very interesting. So, at that time, in addition to our interest in the tokamak, which I first suggested, we were doing in parallel work with Cohn and Halverson on laser produced plasmas.

By the way, on the X-ray laser, the work that Dan Cohn and I did was picked up by Princeton. Remember, we created this hot plasma filament, and we heated the plasma, but we never studied the temperature of the plasma. We should have, but I guess the support for that program ultimately was discontinued by the Air Force.

But Princeton reproduced that work, and used that filament to create X-ray laser interaction. Because, it turns out, you heat the plasma, and after you turn off the pulse it's adiabatically cooled, and there's a mechanism for charge exchange or something to invert.

We heard that these things were in the air at the time, but I never realized that our system would produce that kind of phenomenon. They claimed they made an X-ray laser ... or at least they saw stimulated emission of X-rays. How true it was I don't know, but I know they were using that phenomenon that we developed to produce a hot filamentary plasma, which is something we would have studied anyway. We were studying X-ray emission from those things, but then the program was canceled and we had to drop it.

Synchrotron radiation linewidth too broad for an X-ray laser

The synchrotron radiation is too broad. We did the calculation. I don't know whether I published it. I came to the conclusion that we couldn't make an X-ray laser out of synchrotron radiation, even though it radiates in the X-rays, because the linewidth is so broad. I do remember what the curve looks like.

Magneplane program

Another program that got started was also with Henry Kolm. He and Richard Thornton [8.46] from the MIT Department of Electrical Engineering worked on the development of the Magneplane. This was supported by a Sloan grant, and also was a collaborative effort with Raytheon. It actually was demonstrated very dramatically at Raytheon some years later. I think it was the pioneering effort in magnetic levitation as a practical application.

Nevertheless, there was always a suspicion, and the sponsors as well as our review committees always questioned whether it was appropriate to use NSF money as seed money to get some of these programs started. I don't think we did that, but even if we did I thought it was appropriate. Today I don't think it would be questioned. But at that time that was a problem that often came up in our annual review.

Hybrid magnet

During this period we also began working earnestly, with the insistence of our sponsors and review committees, on the hybrid magnet, which we had conceived in the early 1960s. Now, in the early 1970s, we were going to add a superconducting magnet, which would add about 40 kilogauss to a 200 kilogauss water-cooled magnet. For one thing, we could reduce the power required to go above 200 kilogauss. Instead of using all four generators, we could use two. Or, with all four generators, go at least to a quarter million gauss. I think this was the goal.

Alcator, cryogenically cooled magnets

At the same time, we were also improving our designs and development of Bitter magnets, some for the hybrid magnet. We were experimenting with different designs which outperformed the original Bitter designs. We built a long pulse cryogenic magnet for the Air Force. This cryogenic magnet provided some experience for Alcator, which was going to be cryogenically cooled.

So we got into cryogenic magnets first for the Air Force, but ultimately that was the concept that the Alcator evolved into, although my original concept was to use a small water-cooled magnet just to demonstrate the improved capability of Alcator.

But the use of cryogenic systems was the idea that that evolved into with Bruce Montgomery and coworkers.

Visit to Lumomics in 1973 leads to nonlinear mixing to produce laser lines in the far infrared

Our lasers were sort of homemade. When Lumonics came out with their commercial lasers, I decided I'd like to visit them and find out more about their performance and cost and what we could do.

When I went there, I discovered that they had put a grating into the laser, so that they were able to tune in two rotational-vibrational bands, the 9.6-micron and the 10.6-micron bands. As they varied the position of the grating, which was inside the resonant cavity, and sometimes formed part of the cavity, they were able to step tune between various rotational lines covering the 9- to, I think, the 11-micron region.

It dawned on me, at that time, that we could mix this in a nonlinear crystal and obtain infrared radiation from 100 microns well into the microwave, almost to the microwave, certainly into the millimeter. I came home with this idea and discussed it with Roshi Aggarwal, and he thought it was a good idea. He said that there were some nonlinear phenomena observed in things like gallium arsenide, cadmium telluride.

So I decided to investigate the thing theoretically, and we ended up using the non-collinear matching technique. Aggarwal built a very nice system with mechanical devices that made it possible. And he and I obtained support from ONR, as well as some support from the National Science Foundation.

We produced hundreds of lines. In fact, we calculated we could have many more hundreds of lines from 70 microns well into the millimeter, around 2 or 3 millimeters. And of course we could have gone further. We did it using the pulsed laser.

The theory that I developed, this is the kind of thing I used to do during the summers. This came to be a very interesting device. Later on we did it cw as a master's thesis for Mike Rosenbluh [T43] and he made it work also.

We produced very high powers using this technique inside the gallium arsenide, and we showed you can produce all these lines.

Laser plasmas

Dan Cohn and I began to do a theoretical study of interactions of plasmas with submillimeter radiation. High-power submillimeter lasers, optically pumped with CO_2 lasers, became available. Dan Cohn decided this was a good tool, and would make a good combination for studying the interactions of these submillimeter lasers with plasmas.

I suggested to him we ought to do the analog of my thesis, in doing cyclotron resonance breakdown. He and I worked out the theory, and we determined it was feasible. Ultimately there were two theses that investigated this.

Later on, with the help of Temkin, Dan made a very nice study. In fact, in the back of my mind was the thought of using cyclotron resonance heating, which was something I dreamt about for years. This was a modest beginning. Indeed, we got some nonlinear phenomena which Temkin and his students analyzed. But we never did study the heating effects, but I'm sure they occurred.

Han Le, quantum wells, Auger effect

Around the mid-1980s, I had a contract with ONR through George Wright [8.20] to study luminescence of quantum wells in a magnetic field under high intensities. Han Le, a postdoc, undertook to do these experiments.

We observed a very interesting phenomenon. As we increased the intensity of the excitation, we observed an additional line, which we dubbed a new phenomenon, but later we decided it was a plasma phenomenon. We observed a line below the exciton line. We later discovered that there was some theoretical work. Apparently, it is some sort of plasma effect which produces a collective state below the exciton, at densities that correspond to the levels of excitation when you get a laser.

So, in other words, the lasing transition may be below the gap in many of these materials. And in the example we're just doing, this is in 1998, where we're looking at quantum wells and luminescence, it is possible that – I don't know the level of excitation, I'll have to check with George Turner – but certainly the lasing may take place at a longer wavelength than the energy gap. Consequently the theory, which doesn't take this effect into account, appears at a shorter wavelength by about 10 percent, and the difference between the experiment and the theory may be accounted for by this high intensity excitation and states from below the gap. So we really don't know … at the moment I don't think the theory corrects for this. So I'm not surprised therefore that there is a 10 percent discrepancy between theory and experiment.

Han Le's thesis [T30] was on the motional Stark effect and related problems, but his postdoctoral work was on semiconductors, and this experience and these studies prepared him very well for the work he has been doing at Lincoln Laboratory on quantum well semiconducting lasers. So it looks like that we prepared him well for the real world.

He's been at Lincoln about 15 years. I think his program was one of the last projects I directed.

The other thing that Han Le has done here, which goes back to earlier work that we were concerned with in these mid-IR lasers, is the inter-valley absorption, which he believes is the major culprit for limiting the efficiency of these lasers in these type II indium-arsenide gallium-antimonide quantum wells. And this is the work that we're going to be concentrating on, both theoretically and experimentally, to see if we can minimize these.

Most people attribute the losses to Auger effect. But Han Le believes that the inter-valence band transitions, particularly the light to heavy hole, is the major loss mechanism. And this can be very easily investigated. In fact, by doing magnetic experiments on these losses in emission, which is what I'm proposing, you can learn a lot more about this mechanism and how to minimize it. So the magnetic experiments, although at the moment I thought it was going to be only of academic interest, may be of practical interest in making us understand how to treat these losses and how to avoid them.

Uranium isotope separation with high-power lasers

About that time, in the mid-1970s, the big problem that arose in the laser sciences, both with the Russians and the United States, was the question of using high-power lasers for isotope separation. This was separating uranium isotopes, I forget which isotope. Usually you did it in a gas with UF_6. In some related experiments they did it with SF_6.

It turned out that one of the transitions was in the 8 micron region, and we proposed a four photon mixing of two CO_2 lasers in germanium, and by doing phase matching very similar to what we did in the far infrared, except this was in a third-order nonlinearity … four photon mixing.

We were very successful. We convinced Los Alamos – I forget the guy's name, an MIT graduate – that we could build such a device and they could use it. We got something like $100,000 a year for two years to do the task. Indeed, we built the device, sent it to them. They never used it. But the Russians did that particular experiment at that wavelength, and reported at one of the conferences that were held yearly at one of these ski resorts, Snowbird in Utah.

Optically pumped lasers

About this time, this was in the 1970s, we got very interested in optically pumped lasers for plasma studies.

In fact, at that time, Dan Cohn and I argued about which way to go, whether we should use the nonlinear mixing or optically pumped. But by that time there were so many gases that had different lines, all the way to the submillimeter region. It turns out that with optical pumping we could get – using CO_2 as the pump – we could get a lot more radiation lines in the far infrared than with nonlinear mixing. We could get more than before, and get more power than we could with nonlinear mixing.

So we got into that game. I think that was about the time Temkin joined us. We began to do the physics of the optically pumped lasers, and we even got into the theory of the optically pumped lasers.

Submillimeter lasers, laser-produced plasmas

As I said, at that time we were very much interested in these submillimeter lasers, and also in some theoretical work on it. Also, we were doing a lot of work on laser-produced plasmas.

When the Air Force first took us over in the 1960s, they didn't want us to do any plasma fusion work. So we never did much work on it until the NSF took us over, which allowed us to go after other sponsors.

Through the help of Art Guenther [8.44], and I guess there was a colonel or major who was in charge at AFOSR, I was able to get I think about $300,000 a year to do CO_2 laser interactions in a magnetic field in plasmas.

This is the work that Dan Cohn and Ward Halverson were doing. That really got us started in this work. We also did some very interesting theoretical work with Sunny. I mean, these were related. Dan and I worked on the theory of using high-power submillimeter lasers for cyclotron resonance breakdown, which turned out to be a thesis project for two of our students; Hacker [T21] I think was the principal one.

Fractional Quantum Hall Effect discovered by Bell Labs visitors at the Magnet Lab, merits 1998 Nobel Prize

One of the interesting pieces of research that began here by one of our visitors, Esaki and coworkers from IBM, was the study of the superlattices in semiconductors in a magnetic field. Tsui of Bell Labs, who ultimately got the Nobel Prize, also was studying high magnetic field effects in these two-dimensional systems, inversion layers [8.47]. So, that was the beginning of these visitors who were studying these two-dimensional quantum systems.

Motional Stark effect in a magnetic field, Dave Larsen + Ben's theory

We were studying the motional Stark effect in a magnetic field. This must have been the time that we developed the theory that Dave Larsen and I had, the complete theory with my analytical results, to explain the motional Stark effect. We had very nice theoretical fits of the line shapes to explain the phenomenon.

Submillimeter laser for Thomson scattering as a plasma diagnostic tool

I think one of the really new exciting pieces of work that began [*around 1978*] was the development of high-power submillimeter lasers for Thomson scattering as a plasma diagnostic tool. This was the work of a group of us, with Dan Cohn leading it, and with some theoretical work done by Woskoboinikow, Praddaude [8.48], Dan Cohn, myself, and Bill Mulligan.

We began an important collaboration with Fetterman and Tannenwald at Lincoln for the detector that we had to use, the mixer. We essentially were building a laser radar system. We were scattering from the plasma, and then we had a local oscillator, and we mixed it in a heterodyne detection system. The detector was actually provided by Fetterman. An elaborate system. Praddaude, after he came on board, played a dominant role here, both in the subsequent experimentation and the theoretical analysis.

Magneto-optics expands into ternary compounds

We continued our work on magneto-optical studies. We looked at ternary compounds. This was the work of Steve Groves [*1934–2006*], who must have grown them at Lincoln, and then did the magneto-optical studies and measured the masses and parameters, which today are very important for many of the things we're growing. They confirmed experimentally the estimated properties of these things, as you extrapolated from binary compounds.

Atomic spectroscopy with Terry Miller of Bell Labs; Ben's theory ("one of the cleverest things" he did)

We were doing some very interesting atomic spectroscopy about this time with Terry A. Miller at Bell Laboratories. Panock, Rosenbluh, and I were working with them. And this comprised the theses of Panock [T25] and Rosenbluh [T23].

Using the theory that we developed in that paper that I published with them, we did very sophisticated theoretical calculations of the Zeeman effect of these Rydberg states in a magnetic field. That comprised Panock's thesis [T25].

The techniques we used were the ones that were developed by Rosenbluh, and the analysis of the line shapes involved the theory that I developed for them. They put it on a computer. It was amazing how, by varying the parameters on the computer, they would fit this peculiar line shape.

That involves the theory that I developed, and it explained these complicated line shapes. It was one of the cleverest things I did!

Dave Larsen started, but his analysis was computer driven and was very clumsy in doing the integral. But I got a simple analytical expression, and then entered the phenomenological broadening on top of it, and then the kids could do the computing very conveniently. I got an analytical expression that explained the phenomenon. That's the sort of thing I love to do.

National Magnet Laboratory, just a "magnet factory?"

This business of just pushing the magnet technology for its own sake, and spending so much money on it, I mean, they began to push the Magnet Lab toward that. It became more important than the research. I think this was absolutely stupid. That's not good scientific leadership. But this is what committees do, in the NSF and sponsors. They had no business directing research. It was a waste.

That's what they remember toward the end. We were labeled a "magnet factory" instead of a magnet research lab. And that's what got us into trouble [8.49].

CHAPTER 9

Professor of Physics at MIT, 1965–1981

Relinquishes role as Associate Director of Lincoln Laboratory, joins MIT Physics Faculty

In 1965 I decided, at some prodding by the Air Force, who said I couldn't continue being Associate Director of Lincoln Laboratory and Director of the Magnet Laboratory, that I had to make a choice. I decided that I would leave Lincoln Laboratory and become a professor at MIT [*and continue on as Director of the Magnet Lab*].

I discussed it with Charlie Townes [7.69], who was the MIT Provost, and he was all for it. So was John Slater [5.5]. Both were enthusiastic about this. Because John Slater wanted me to come to the campus, and because I had a research program on the campus at the Magnet Laboratory, it was the appropriate time.

In addition, I got tired of being Associate Director of Lincoln Laboratory because, between that and the Magnet Lab, I spent most of my time in administration, and I began to drift away from my science. This disturbed me a great deal because I had a lot of good ideas still.

I also had an offer, about that time, from Duke University to come down there. They offered me a chair if I would come down. Gordy [9.1] was after me for years.

So I told Charlie Townes, around 1965, after they were dragging their feet, either he makes me an offer or I go to Duke. Pretty soon they went through the papers, and I got an offer from MIT, about the same kind of offer in terms of salary (no chair) that I would have gotten at Duke. Of course, this was a lower salary than I got at Lincoln, but with summer pay and consulting, I made about the same amount of money throughout the year as I did as Associate Director of Lincoln Laboratory. So I didn't suffer financially.

What I realized and wondered about, was that being a professor at MIT meant that I had to teach, direct students, and also run the Magnet Laboratory. Of course, I was already running two laboratories. When I told this to my Rabbi, he used the phrase, the Hebrew phrase, I hope you have the *koach*, which means strength in Hebrew, to do both of these. Apparently, even when I got promoted to Associate Director of Lincoln Laboratory, he wondered whether I could cope with it, and I did, except I wasn't happy doing full-time administration.

But it did work out. The amazing thing is, as professor, right after I got in, I taught every term, which wasn't necessary. But I taught every term at the beginning, both graduate and undergraduate courses, and also directed students. In fact, I acquired more students, ran the Magnet Laboratory, and did science.

I often wonder where I got the stamina and interest. I probably neglected my family, because I would go home and prepare my lectures at night, after the children went to bed. But I managed somehow.

During the summer, I was able to do much better because I didn't have to teach, so I could do more theory. I noticed that the number of projects I started or engaged in during the summer were more active, or I was more active in these projects at the Magnet Lab.

Appointed to the MIT Physics Faculty

You can't get a faculty appointment at the MIT Physics Department without having extensive publications, good ones that are rated highly by outside and inside reviewers. That's why I had no trouble at all when I got into the Physics Department.

In fact, I told Charlie Townes, who was anxious to get me there when I made my decision to leave Lincoln Laboratory and resign as the Associate Director, and they were dawdling, I told him I got an offer from Duke. Gordy [9.1] was after me three times: when I left the Radiation Lab, and later when I was at Lincoln Laboratory, and once again when I was ready to retire.

In fact, I went for an interview there, and I was considering it seriously. He gave me a good offer. About the same offer that MIT gave me a little later. So, I told Charlie Townes that they were going to give me a chair at Duke. Then they immediately acted.

When I came up for review, it turned out that they discovered I had more publication references that anybody in the Physics Department. I didn't find this out until George Benedek [9.2] told me. So that was a surprise, which meant my work was at least appreciated.

I probably had very good recommendations from a variety of people. And, of course, I had the Buckley Prize. So I had no trouble getting the appointment. Bill Buckner, who was the head of the MIT Physics Department, said, "Well, Ben Lax, you know, may not be teaching, but he'll be an asset."

FIGURE 9.1 Prof. Benjamin Lax. (Photo credit: Fabian Bachrach.)

On being a good teacher

It turned out I wasn't a bad teacher either, when I concentrated on it. So they figured I was primarily a researcher and an administrator, but I could do all of these other things just as well when I set my mind to it.

I was always a good lecturer. But even in lecturing and teaching, the secret of success is dedication and preparation.

I used to attend all the lectures, even in freshman physics, all the lectures of the lecturers when I taught the recitation section. I would of course do all the problems in advance of the students. I found out that's the way you help them most.

So you put yourself into their shoes, and recall when you were just a student yourself. But as a recitation section teacher, I probably was more conscientious than most of the students, just as I was when I was a student myself.

And so, in fact, they liked my recitation section. I had pretty good attendance for the most part. The ones that didn't attend didn't do as well because I had a scheme. I wouldn't look at the actual exams prior to their being made up. But I would give them difficult problems that they might get on the exam. And I would guess about half of the problems, if not in exact detail but in principle, half of the problems that they got on these exams, particularly on the final. I guess this was very helpful to the students. But that was perfectly legitimate.

Magneto-plasma experiments

George Wright [8.20] was the MIT graduate student of Arthur von Hippel, but George turned out doing his thesis [T55] at the Magnet Lab with me.

George Wright knew what he wanted to do, but what he wanted to do was impossible. He wanted to do cyclotron resonance and magneto-optical effects in mercury-selenide, mercury-telluride, but the materials were so impure you couldn't do either. They were so heavily doped.

So one day, while he was talking to me, I said to him, you know what, I've got an idea. I got this idea for a magneto-plasma phenomenon. That's the analog of what Appleton and Hartree tried to do in the ionosphere. They looked at reflection, and they barely saw it split, the magneto-plasma edge.

I went right home that evening and worked out the theory, and predicted that the plasma edge would be split by the cyclotron frequency. In fact, that paper [P9.1] appears in many textbooks. That became his thesis [T55]. He did a complete job, did a detailed analysis, did lots of experiments [P9.2].

In two weeks we showed the phenomenon to occur in indium antimonide, and then he ultimately was able to do it in mercury telluride and mercury selenide. So it did work, and we were able to determine the effective mass of electrons in mercury telluride, I think it was, with n-type doping.

George Wright went on to Lincoln Laboratory, and became an assistant group leader there. He worked with Mooradian. He and Mooradian did the experiment on observing the Raman scattering by plasmons in gallium arsenide, which was doped with different doping concentrations, and they showed that there was a plasma-phonon interaction [9.3].

Mooradian and Wright [9.3] got the typical anti-crossing of the plasma-phonon, and that's what gave me the idea for Dick Stimets' thesis [T5] for doing the magneto-plasma experiment with just one doping. You have a doping that's near the anti-cross point. You put on a magnetic field, and you can make it either phonon-like or plasmon-like by just tuning the magnetic field.

And that was Stimets' thesis. He and I did the theory, and that was a little more complicated than the plasma. It was essentially the same, in fact we were doing it with George Wright's thesis [T55], except you did it in a region where the plasma frequency and the phonon frequency were equal, and then they coupled strongly.

In other words, the energy is equal to the square root of the plasma frequency squared plus the phonon frequency squared. You could work this out theoretically. You could do it classically or you could do it quantum mechanically. And remember, Rabitz and I did the quantum theory of this phenomenon [P9.3]. It was very interesting.

Coaches Tim Hart in oral qualifying exam

Tim Hart [9.4] was one of the brightest students I ever had. In fact, he knew things very well, but he was emotionally very fragile. He'd get very scared at written and oral exams. He'd freeze up. Now he didn't do too badly on the written, but on the orals, he'd just go to pieces.

So he came to me, and I think he had flunked the exam once or something like that. What I did was work with him for a year. Gave him typical exams, told him the key to any exam, oral or otherwise, is just to study the way I used to study for my exams. He did that and then I gave him trial exams, and I told him I'm going to ask you one of the same questions that I asked in one of these trial exams, maybe with some variation, but I'll be the first one to ask questions. Once you settle down, you'll be fine.

He passed it. He turned out to be one of my best students, and did one of the best theses with me. In fact, I was at that time teaching the many-body theory of solid state, and we used what I learned to explain his thesis data.

When Tim Hart was ready to do his thesis, had passed his oral and written qualifying exams, Roshi Aggarwal and I decided he was going to do Raman scattering in silicon. It's a very classic experiment and theory. That turned out to be a nice thesis [T8] and application of the second quantization of phonons, and it's really a classic problem.

Tim and Roshi did the experiment and I worked out the theory, and then we found out there was a man in Canada who did a similar theory [9.5]. So between myself and that reference, he was able to work out the complete theory, and everything … theory and experiment … agreed very well.

When you do Raman scattering, you shoot in light and you excite an optical phonon, and you measure the frequency of the line, of the shifted line, on a spectrometer. And so you have a certain linewidth, which of course varies with temperature.

Now the question is, what determines the linewidth? Well, I figured out it was the decay of the optical phonon into two acoustic phonons, and you can represent this by quantization of both, and with a third-order term, and you do some statistics, as a function of temperature, and you've got the linewidth. And theory and experiment agreed, also the frequency shift, the two were related.

Raman scattering in europium chalcogenides

It was about that time, I think, that we handed over to Millie Dresselhaus the Raman setup that Roshi Aggarwal had built, in which we had done theses with Tim Hart and I guess others. She began using that setup to look at the Raman scattering in the europium chalcogenides. This is a magnetic material.

Safran [T22], who wanted to do a theoretical thesis with me, got interested. We got him interested in the problem. He, with the help of Gene Dresselhaus, interpreted some of these transitions that we saw as this material went through several magnetic transitions as a function of temperature, from ferrimagnetic to anti-ferrimagnetic to different things. What was interesting was, it would change essentially the symmetry.

In other words, the magnetic structure, or the magnetic crystal, had a different size than the actual crystal. With each magnetic change you would see additional phonon lines. In other words, the original Brillouin zone was folded. We figured this out, and Safran [T22] did the details, and he accounted for the change as well as other things.

Nonlinear magneto-absorption

About that time, we also got very interested in doing nonlinear magneto-absorption in semiconductors where we saw a lot of transitions. This was the thesis work of Favrot [T58] with Roshi Aggarwal and myself.

Spin-flip Raman laser in high magnetic field

Sunny Yuen became an MIT graduate student about the same time as Neville Lee. Roshi and I got very interested in doing a high-power tuning of the spin-flip Raman laser in a high magnetic field. We did a very nice experiment and got a broad band tuning. Then Sunny Yuen got very interested in doing the density matrix treatment of this.

I tried to talk her out of it because it was much more complicated, even though we had done some interesting work in that regard. In fact, Peter Wolff [7.47] and I both tried to discourage her, but she went ahead and solved the problem. It was a rather complicated problem, and I think this was the first such solution. It was a nice thesis [T15].

Magneto-optics of mercury cadmium telluride

This was the time that Margaret Weiler [9.6] did a very comprehensive study of the magnetoreflection in mercury cadmium telluride, both experimentally and theoretically. This was part of her thesis [T20]. It was I think one of the most sophisticated, first-class works of anyone in this field.

Margaret deserved lots of credit for being a first-rate physicist. In my opinion, she was one of the best theorists in this field of magneto-optics, and that included Laura Roth and the others who were working in there, including the outside people at NRL.

Ben, a novice skier, does giant slalom course in Innsbruck

During the 1970s, when we were doing our plasma research and attending plasma physics conferences, at the age of 57 I learned to ski.

These winter Gordon conferences where we presented our plasma papers were held at all sorts of ski places, like Snowbird (Utah) and Steamboat Springs (Colorado). About this time, I think it was 1973, I went to a conference in Alta, a ski resort in Utah. I was watching them ski, and I thought, this doesn't look hard. So I borrowed some ski clothes, got into skis, and

FIGURE 9.2 (L-to-R) Prof. Benjamin Lax and his MIT colleagues, Prof. Herman Feshbach, Prof. Francis E. Low, and Dr. Roshan L. Aggarwal, at an all-day symposium on June 15, 1981, organized by the MIT Physics Department to honor Lax's retirement as Founding Director of the Francis Bitter National Magnet Laboratory. (Courtesy MIT Museum.)

I found myself skiing very easily. You know, I was athletic, and was in good shape in those days, at the age of 57.

So then, when my older son Daniel heard that I had skied, he took me to Blue Hills Ski Area (in nearby Canton, Massachusetts), to the beginner's slope, and I learned to parallel ski there. Next he took me to Waterville Valley, a ski resort in New Hampshire. It was 1973, winter of 1973, or January or February of 1974, and I took lessons there. I took to it very quickly. And that's how I became a skier.

It was the following year, 1975, that Dan Cohn and I went to Innsbruck, Austria, to a plasma conference [9.7] to present papers. We decided to go skiing there. We went to the Aximer Lizum Ski Resort. We went up to the top, and then we traversed to go ski on the intermediate slope. You have to traverse all the way.

So we started skiing on the intermediate slope, and we got completely fogged in. We didn't know how we were going to get back to the traverse to go down on the gondola. We didn't think we could ski down because we didn't know the terrain, but we had no choice.

So we decided to go. You know, up in Austria you can have crevasses and everything. We managed to end up at the place where they were going to have the Winter Olympics the next year [9.8]. We ended up on the giant slalom course! Here I was, I had skied only one year, the year before when my son Daniel taught me, and I decided I wasn't going to take off my skis. So we skied, slowly, on the moguls, and managed to get down safely [9.9]. So that was my great adventure in Austria. Dan Cohn and I skied quite a bit there, and then we presented our paper [P9.4].

CHAPTER 10

Emeritus Years and Consulting, 1981–2006

Consults at Lincoln Laboratory and Raytheon

I'd been a consultant to Lincoln Laboratory all the time I was a professor at MIT. But just about six months before I was ready to retire, I was consulting both for Lincoln Laboratory and Raytheon. One of the projects that Raytheon wrote a proposal for was an interband circulator to be used in the CO_2 laser radar. Well, I'm the inventor of the interband circulator.

Invents the interband circulator

That came about many years before, when Dick Brown did his thesis [T54] in 1958. He was looking at the Faraday rotation in semiconductors as a function of carrier density, and also as a function of wavelength. As he went to shorter wavelength, approaching the energy gap of the semiconductor, I think it was indium antimonide, it turned out that the sign of the Faraday rotation changed from positive to negative, or the other way around.

Dick Brown didn't know what it was. But I concluded that, in addition to the free carrier rotation, there was contribution due to the interband transition, even though we were below the energy gap. But, as we approached the gap, it becomes resonant. It was a virtual transition. Later on, I developed a theory for this [P10.1]. I concluded that this was the interband Faraday rotation. That was one of the first discoveries of the interband rotation.

When Nishina joined me in 1960 at the Magnet Lab, he did some experiments on the interband rotation, which we reported subsequently at a conference at GE Laboratory in Schenectady [P8.1, P10.1, P10.2].

I had a number of students and postdocs working on the theory. In fact, one of the Polish postdocs, Kolodziejczak, worked with me on the theory [P10.3]. But I fundamentally developed it, and Mavroides and I reported this in a review paper [10.4] in great detail. Halpern did his thesis [T2] on this also.

When I was consulting for Lincoln Laboratory, I think in the 1970s, Kingston [7.71] wanted to build a radar, and use the free carrier Faraday rotation for the isolator for the CO_2 laser radar. I said to him, "That's silly. You can do much better with the interband because the losses are less and the figure of merit is much higher." He didn't pay any attention to me.

Consults for Lawrence Livermore on interband Faraday rotation isolator

But then, at a cocktail party some months later at one of the meetings, this fellow from Lawrence Livermore came to me and says, "Professor Lax, you published a lot of papers on the Faraday rotation in semiconductors. I need an isolator, and I want to use the Faraday rotation, so I was thinking of using free carriers."

I said, "We've been through this. You don't want to use the free carriers. You want to use the interband." He wanted to use indium antimonide, and I said, "That's no good, for the simple reason that, with the intensities you're using at the CO_2 laser ... they were using this for fusion ... with the intensities you're using, you're going to get multiphoton absorption across such a small gap. Why don't we use something with a higher gap?"

I suggested germanium and cadmium telluride, except we hadn't measured or studied the Faraday rotation in those materials. But I said, "I have the theory and I have the various parameters, the g-factors, and I can calculate it."

He asked me to consult for them. So I went down there, spent two days there. We looked for the Verdet constant of germanium and cadmium telluride. It wasn't in the tables. So I sat down and calculated and predicted what it was. Subsequently he measured it, and I had guessed the answer to within 10 percent. Laura Roth and I had done quite a bit of calculations.

She did the calculation taking into account the g-factors of both the valence and conduction bands. So I was able to figure it out.

That was the first application of the interband Faraday rotation. He used, I think, germanium, which has a higher gap. The direct transition was 0.8 eV, not 0.2 eV like it is in indium antimonide. And it worked. I was credited with the invention. I didn't take out a patent on it, but subsequently such an isolator was built at Lincoln when they started building their Firepond CO_2 laser radar, and I think also for our work that we did at Livermore.

The interesting thing about Livermore was, after we did the successful experiment, the man who was in charge of the laser fusion effort at Livermore, when we wanted to publish the paper, said to my coworker, "We don't have to include his name on this paper. He's a consultant." When I heard that, I called him up and I said, "The hell with you. Take your money. I want credit for this. I want to be a coauthor on this." I got both the money and the credit.

Subsequently it was built here at Lincoln Laboratory, with indium antimonide, for the CO_2 laser, because here they were using it for radar, not for fusion where it was much higher power and the laser was focused. So it worked out [P10.4].

After I retired, or just before I retired, in 1981, I was consulting for Raytheon, as I mentioned. But Lincoln Laboratory asked them to submit a proposal for building and designing an isolator based on the interband effect for the SDI program, because Firepond was being revived and they were going to have a laser radar there for looking at missiles.

I was working on it at Raytheon, and then, just before I retired, Bill Keicher said, "Why don't you work on it with us for our program, since you're also consulting for Lincoln." And I said, "I can do that. Of course, I may have a conflict of interest." So I stopped working at Raytheon on it, and started working on it for Lincoln. Then he offered me a job as a senior scientist at Lincoln a few months before I retired. I thought that was what I was going to do.

John Deutch [5.10], who was the MIT Provost at the time, just 10 days before I was ready to retire, said, "We can't rehire, we can't hire a retired MIT professor at Lincoln Laboratory." So we scratched our heads, and Lincoln Laboratory proposed that I be a subcontractor. So I became a subcontractor and a consultant for the lab.

In fact, Dan Cohn also was a consultant for Lincoln, so the two of us worked on the specifications for the isolator. It's really a Faraday rotation isolator, it's not an absorption isolator. But it was the interband isolator that I had invented. So I worked with that.

Develops cadmium telluride modulator for Firepond laser radar

But during that time, it turned out that they wanted the Firepond CO_2 radar to be much more broadband. And the way you do that is to make a modulator of cadmium telluride inside a microwave waveguide. Cadmium telluride is a piezo-electric material, it's a nonlinear material.

The proposition was that you were going to mix the laser light, the CO_2 beam, with microwaves inside this microwave waveguide with a micro-wave traveling wave tube, which you can tune. So this way you modulate the output beam. Of course, the output beam is shifted by the microwave energy plus the modulation frequency, which is a fraction of that. So we worked on that. It's an electro-optic material, it's an electro-optical effect, and we worked on that.

But the problem is, as you vary the microwave frequency and shift the laser frequency coming out of the modulator, it has a sine x over x variation, and the half-width, instead of being say 2 gigahertz, is of the order of a half-gigahertz. So that was inadequate.

So I got the idea, and apparently two of us got the idea, except the other guy didn't know how to do the theory. I got the idea that you slice up the crystal. You make the crystal long enough so that you get essentially 100 percent conversion. I showed that if it's about 20 centimeters long, it converts the entire CO_2 laser to the modulated frequency. The required length depends on the microwave power. For the kind of traveling-wave tubes we have, it turned out to be about 20 centimeters.

So I suggested we cut up the crystal and, as the microwave and the optical beam get out of phase, we use another microwave tube to readjust the phase. And I used a variational parameter as the adjusted phase and worked it out for two sections, three sections, and then I realized as I was doing it that the coefficients that I was writing out for three layers and four layers were the binomial coefficients. So I was able to write out the result after that for any number of sections as a summation. It was a very nice mathematical problem.

It worked. That was the first time I got into this. That was one of the things that I worked on here at Lincoln Laboratory. It was very successful and I published the paper [P10.5]. Richard Eng did the experiments, and Neville Harris built the actual device. It worked very well.

But the other thing that happened is that, when you transfer … when you shift the frequency up you want to shift it down, and they were planning to shift it down without doing anything with the downshift. And I showed, if you make a modulator for both the upshift and downshift, you increase

FIGURE 10.1   Blossom and Ben Lax in Lascaux, near Bordeaux, France, May 14, 1993. (Photograph by Prof. John David Jackson, courtesy AIP Emilio Segrè Visual Archives, Jackson Collection.)

the bandwidth, and that reduces the number of sections you had to cut the crystal in each modulator. That was the ultimate invention.

That was one of my important contributions to the Firepond CO_2 laser radar. This took a space of about two years, theory and experiments. We got this additional idea, the up and down modulator. It was a very nice piece of work, and I published with my coauthors [P10.6, P10.7]. The computations were done by a couple of other people, Richard Eng and Charles Summers. That was a very successful effort.

Reflection isolator

Then, for some reason or other, I got interested subsequently here at Lincoln in millimeters. I don't know how that came about. And, not just millimeters, it could be with the CO_2 laser. I got the idea that, instead of transmission isolator, we can make a reflection isolator. For millimeters with ferrites this would be a great device.

I got the idea from Gerry Dionne who was in another group where they had a millimeter radar. They were using a ferrite, a transmission isolator, a Faraday rotation isolator. But as the microwave power became bigger, they cooled it on the edge of a cylindrical slab of ferrite. And that is not an efficient way of cooling.

I got the idea that you could use a reflection isolator. This turned out to be even a more complicated problem, because the problem was that you have to do the reflection at oblique incidence. So your beam, which is polarized, which could be s and p polarization, hits this slab, which not only has a dielectric tensor in a magnetic field, but it's also gyrotropic.

It's very complicated. I worked out the theory based on some of the work of Joshua Zak [10.2]. Richard Eng and Neville Harris did the experiments. In fact, we got a magnet from the Magnet Lab. Again, it turned out to be very successful. It was a very nice piece of mathematical, theoretical work, which agreed pretty well with experiment [P10.8, P10.9].

Quarter wave plates on the ferrite

That's when I got interested in building quarter wave plates on the ferrite, because when the microwave hits it, you don't want it to be reflected. You wanted it to be matched. So you use a dielectric. Or, the idea I got was, since these were of millimeter dimensions, you could take the dielectric and dice it either into slits like a grating or into squares so the effective

FIGURE 10.2 Prof. Benjamin Lax, 2005. (Reprinted with permission. Courtesy of MIT Lincoln Laboratory, Lexington, Massachusetts.)

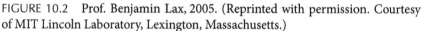

dielectric constant of the layer is the intermediate. In other words, you could do it on the ferrite itself.

That's when I got interested in the theory of dielectric coatings. I read MacLeod's book [10.3], and did some of my own theoretical work. One of the interesting things that MacLeod said in his book was that, with the quarter wave plate, you can't match both the *S* and *P* polarizations. But I showed if you have multiple layers, at least two layers, you can perfectly match at a particular angle, both *S* and *P*. And that was an interesting sidelight.

This is why I'm saying now, here at Lincoln Laboratory, we're doing a lot of dielectric coatings on the semiconducting laser. I've gotten interested again in the dielectric layers, but now they're more complicated because we're trying to make three-layer coatings, and that gives us more flexibility for making broad band and also angular independence.

Celebratory symposia, final publications

[On June 15, 1981 there was an all-day symposium at MIT to honor Prof. Benjamin Lax on the occasion of his retirement as Founding Director of the Francis Bitter National Magnet Laboratory. Colleagues and students of Ben's spoke on a variety of topics in solid state physics and plasma physics.]

[On May 23, 1986 a half-day symposium was held at the Magnet Lab to celebrate Ben's seventieth birthday. Speakers included Nobel Laureates Leo Esaki of IBM and Nicolaas Bloembergen of Harvard, as well as C. Kumar N. Patel of Bell Laboratories, and Alan McWhorter and Paul L. Kelley, both of Lincoln Laboratory.]

[Ben's last publications were in 2005. One was an article in the Lincoln Laboratory Journal [P10.10]. The other, appropriately, was an invited paper in a special session commemorating the fiftieth anniversary of cyclotron resonance, which keynoted the 27th International Conference on the Physics of Semiconductors [P10.11]. Other speakers in that session included Gene Dresselhaus, Elias Burstein, Marvin L. Cohen and Klaus von Klitzing.]

Selected Publications of Prof. Benjamin Lax

T HIS SECTION LISTS THOSE publications that were authored or coauthored by Prof. Benjamin Lax and that are referenced in Chapters 1–10. His complete list of publications includes over 300 journal articles and numerous review chapters and books, and is far too lengthy to be part of this printed book. A complete list of all his publications appears online at the CRC Press website associated with this book.

Chapter 5 Graduate School in Physics at MIT, 1946–1949

P5.1 Benjamin Lax, "The Effect of Magnetic Field on the Breakdown of Gases at High Frequencies," PhD Thesis, MIT Department of Physics, August 1949, supervised by Sanborn C. Brown. Available online at the MIT Barton Library website: https://dspace.mit.edu/handle/1721.1/12394

P5.2 Benjamin Lax, W. P. Allis, and Sanborn C. Brown, "The Effect of Magnetic Field on the Breakdown of Gases at Microwave Frequencies," *Journal of Applied Physics* **21** (12), 1297–1304 (December 1950).

Chapter 6 Postdoctoral Work at Air Force Cambridge Research Laboratories, 1949–1951

P6.1 J. R. Terrall and Benjamin Lax, "Perturbation Treatment of Electromagnetic Problems. I. Theory," Abstract I2, Minutes of the 1952 Annual Meeting of the American Physical Society, New York City, January 31–February 2, 1952; *Physical Review* **86**, 595 (1952).

P6.2 Benjamin Lax and J. R. Terrall, "Perturbation Treatment of Electromagnetic Problems. II. Applications," Abstract I3, Minutes of the 1952 Annual Meeting of the American Physical Society,

New York City, January 31–February 2, 1952; *Physical Review* **86**, 595 (1952).

Chapter 7   MIT Lincoln Laboratory, 1951–1965

P7.1 A. D. Berk and B. Lax, "Cavities with Complex Media," Abstract 42.4, *Proceedings of the Institute of Radio Engineers* **41** (3), 427 (March, 1953).

P7.2 B. Lax and A. D. Berk, "Resonance in Cavities with Complex Media," Abstract 42.5, *Proceedings of the Institute of Radio Engineers* **41** (3), 427 (March, 1953).

P7.3 B. Lax and S. F. Neustadter, "Transient Response of a p-n Junction," *Journal of Applied Physics* **25** (9), 1148–1154 (September, 1954).

P7.4 B. Lax, H. J. Zeiger, R. N. Dexter, and E. S. Rosenblum, "Directional Properties of the Cyclotron Resonance in Germanium," *Physical Review* **93**, 1418–1420 (March 15, 1954). Manuscript received January 25, 1954. Letter.

P7.5 R. N. Dexter, H. J. Zeiger, and B. Lax, "Anisotropy of Cyclotron Resonance of Holes in Germanium," *Physical Review* **95**, 557–558 (July 15, 1954), Letter.

P7.6 R.N. Dexter, B. Lax, A. F. Kip, and G. Dresselhaus, "Effective Masses of Electrons in Silicon," *Physical Review* **96**, 222–223 (October 1, 1954), Letter.

P7.7 R. N. Dexter and B. Lax, "Effective Masses of Holes in Silicon," *Physical Review* **96**, 223–224 (October 1, 1954), Letter.

P7.8 B. Lax, H. J. Zeiger, and R. N. Dexter, "Anisotropy of Cyclotron Resonance in Germanium," Invited Paper, *Proceedings of the Third International Conference on the Physics of Semiconductors, ICPS-3, Amsterdam, June 29–July 3, 1954, Physica* **20** (11), 818–828 (December 1954). These proceedings were published as a book by the Netherlands Physical Society in 1954.

P7.9 R. N. Dexter, H. J. Zeiger, and B. Lax, "Cyclotron Resonance Experiments in Silicon and Germanium," *Physical Review* **104**, 637–644 (November 1, 1956).

P7.9a R. N. Dexter and Benjamin Lax, "Microwave Magnetoconductivity Effects in InSb," *Physical Review* **99**, 635 (July, 1955). Abstract L2, American Physical Society Spring Meeting, National Bureau of Standards, Washington, DC, April 28–30, 1955.

P7.10 R. N. Dexter and B. Lax, "Cyclotron Resonance in Bismuth," *Physical Review* **100**, 1216–1218 (November 15, 1955), Letter.

P7.11 B. Lax, K. J. Button, H. J. Zeiger, and L. M. Roth, "Analysis of Cyclotron Absorption in Bismuth," *Physical Review* **102**, 715–721 (May 1, 1956).

P7.12 B. Lax and L. M. Roth, "Propagation and Plasma Oscillation in Semiconductors with Magnetic Fields," *Physical Review* **98**, 548 (April 15, 1955), Letter.

P7.13 R. J. Keyes, S. Zwerdling, S. Foner, H. H. Kolm, and B. Lax, "Infrared Cyclotron Resonance in Bi, InSb, and InAs with High Pulsed Magnetic Fields," *Physical Review* **104**, 1805–1806 (December 15, 1956), Letter.

P7.14 S. Zwerdling, R. J. Keyes, S. Foner, H. H. Kolm, and B. Lax, "Magneto-Band Effects in InAs and InSb in dc and High Pulsed Magnetic Fields," *Physical Review* **104** (6), 1805–1807 (December 15, 1956), Letter.

P7.15 B. Lax, J. G. Mavroides, H. J. Zeiger, and R. J. Keyes, "Cyclotron Resonance in Indium Antimonide at High Magnetic Fields," *Physical Review* **122**, 31–34 (April 1, 1961).

P7.16 S. Zwerdling and B. Lax, "Oscillatory Magneto-Absorption of the Direct Transition in Germanium," *Physical Review* **106**, 51–52 (April 1, 1957). Received December 26, 1956.

P7.17 S. Zwerdling, B. Lax and L. M. Roth, "Oscillatory Magneto-Absorption in Semiconductors," *Physical Review* **108**, 1402–1408 (December 15, 1957).

P7.18 L. M. Roth and B. Lax, "*g*-Factor of Electrons in Germanium," *Physical Review Letters* **3**, 217–219 (1959).

P7.19 L. M. Roth, B. Lax and S. Zwerdling, "Theory of Optical Magneto-Absorption Effects in Semiconductors," *Physical Review* **114**, 90–104 (1959).

P7.20 S. Zwerdling, L. M. Roth, and B. Lax, "Direct Transition Exciton and Fine Structure of the MagnetoAbsorption Spectrum in Germanium," *Physical Review* **109**, 2207–2209 (March 15, 1958), Letter.

P7.21 S. Zwerdling, B. Lax, L. M. Roth, and K. J. Button, "Exciton and Magneto-Absorption of the Direct and Indirect Transitions in Germanium," *Physical Review* **114**, 80–89 (April 1, 1959).

P7.22 B. Lax, J. G. Mavroides, H. J. Zeiger, and R. J. Keyes, "Infrared Magnetoreflection in Bismuth. I. High Fields," *Physical Review Letters* **5**, 241 (September 15, 1960).

P7.23 R. N. Brown, J. G. Mavroides, M. S. Dresselhaus, and B. Lax, "Infrared Magnetoreflection in Bismuth. II. Low Fields," *Physical Review Letters* **5**, 243 (September 15, 1960).

P7.24 B. Lax and K. J. Button, *Microwave Ferrites and Ferromagnetics* (London, McGraw-Hill, 1962).

P7.25 Benjamin Lax, "Cyclotron Resonance and Impurity Levels in Semiconductors" in *Quantum Electronics, A Symposium, edited by Charles H. Townes (Papers and Discussion at the Conference on Quantum Electronics-Resonance Phenomena, Held at Shawanga Lodge, High View, New York on September 14–16, 1959).* Published by Columbia University Press, New York, 1960; pp. 428–449. The title of this paper does not capture what the paper is really about. The real purpose is stated in the paper's first paragraph:

> For some time, semiconductors have been seriously considered as a possible medium for generating infrared and millimeter radiation. Some success has already been attained in generating incoherent radiation in the infrared. Consequently, it is a logical step to consider semiconductors as likely candidates for use as quantum amplifiers and oscillators. A number of proposals have been made in the literature and elsewhere. I would like to review these, comment on them, and also add one or two suggestions of my own. The basic phenomena that are involved in most of these proposals concern cyclotron resonance and impurity levels.

P7.26 A. R. Calawa, R. H. Rediker, B. Lax, and A. L. McWhorter, "Magneto-Tunneling in InSb," *Physical Review Letters* **5**, 55–57 (July 15, 1960).

P7.27 T. M. Quist, R. H. Rediker, R. J. Keyes, W. E. Krag, B. Lax, A. L. McWhorter, and H. J. Zeiger, "Semiconductor MASER of GaAs," *Applied Physics Letters* **1**, 91 (December 1, 1962).

P7.28 A. L. McWhorter, H. J. Zeiger, and B. Lax, "Theory of Semiconductor Maser of GaAs," *Journal of Applied Physics* **34**, 235 (1963).

P7.29 B. Lax, "Magnetospectroscopy in Semiconductors," *Proceedings of the International Conference on Semiconductor Physics, Prague, August 29–September 2, 1960, ICPS-5,* pp. 321–327 (Academic Press, 1960).

Chapter 8 Francis Bitter National Magnet Laboratory, 1958–1981

P8.1 B. Lax and Y. Nishina, "Interband Faraday Rotation in III-V Compounds," Proceedings of the Conference on Semiconducting Compounds, June 14–16, 1961, General Electric Research Laboratory, Schenectady, New York, *Journal of Applied Physics* **32** (10), 2128 (1961).

P8.2 Y. Shapira and B. Lax, "Determination of Effective Masses from Giant Quantum Oscillations in Ultrasonic Absorption," *Physical Review Letters* **12**, 166 (1964).

P8.3 J. Halpern, B. Lax, and Y. Nishina, "Quantum Theory of Interband Faraday and Voigt Effects," *Physical Review* **A134**, 140 (1964).

P8.4 R. L. Aggarwal, L. Rubin, and B. Lax, "Magnetopiezo-Optical Reflection in Germanium," *Physical Review Letters* **17**, 8–10 (July 4, 1966).

P8.5 R. L. Aggarwal, M. D. Zuteck, and B. Lax, "Nonparabolicity of the $L1$ Conduction Band in Germanium from Magnetopiezotransmission Experiments," *Physical Review Letters* **19**, 236–238 (1967); *Erratum* **19**, 1411 (1967).

P8.6 S. Zwerdling, R. J. Keyes, S. Foner, H. H. Kolm, and Benjamin Lax, "Magneto-Band Effects in InAs and InSb in dc and High Pulsed Magnetic Fields," *Physical Review* **104**, 1805–1807 (December 15, 1956).

P8.7 B. Sacks and B. Lax, "Carrier Lifetime and Threshold of $Pb_{1-x}Sn_xTe$ Lasers in a Magnetic Field," *Journal of Quantum Electronics* **QE- 6**, 313 (1970).

P8.8 J. M. Honig and B. Lax, "Performance Characteristics of Thermomagnetic Devices Involving Graded Mass and Gap. I. Generators," *Journal of Applied Physics* **39**, 3549 (1968).

P8.9 B. Lax and A. H. Guenther, "Quantitative Aspects of a Soft X-Ray Laser," *Applied Physics Letters* **21**, 361–363 (November, 1972).

P8.10 S. Y. Yuen, B. Lax, and D. R. Cohn, "Laser Heating of a Magnetically Confined Plasma," *Physics of Fluids* **18**, 829 (1975).

Chapter 9 Professor of Physics at MIT, 1965–1981

P9.1 B. Lax and G. B. Wright, "Magnetoplasma Reflection in Solids," *Physical Review Letters* **4**, 16–18 (1960).

P9.2 G. B. Wright and B. Lax, "Magnetoreflection Experiments in Intermetallics," Proceedings of the Conference on Semiconducting Compounds, June 14–16, 1961, General Electric Research Laboratory, Schenectady, New York; *Journal of Applied Physics* **32** (10), 2113 (1961).

P9.3 R. Rabitz and B. Lax, "Coupled Magnetoplasma and Phonon Systems," *Journal of Physics and Chemistry of Solids* **32**, 359 (1971).

P9.4 B. Lax, D. R. Cohn, W. Halverson, and S. Y. Yuen, "Laser-Plasma Interactions in a Magnetic Field," Abstract Q1 in *Book of Abstracts, Second International Congress on Waves and Instabilities in Plasmas, March 17–21, 1975, Innsbruck, Austria* (Institute of Theoretical Physics, University of Innsbruck, Innsbruck, Austria, 1975).

Chapter 10   Emeritus Years and Consulting, 1981–2006

P10.1 B. Lax and Y. Nishina, "Theory of Interband Faraday Rotation in Semiconductors," *Physical Review Letters* **6**, 464–467 (1961).

P10.2 Y. Nishina, J. Kolodziejczak, and B. Lax, "Oscillatory Interband Faraday Rotation and Voigt Effects in Semiconductors," *Physical Review Letters* **9**, 55–57 (1962).

P10.3 B. Lax, J. E. Mavroides, and J. Kolodziejczak, "Dispersion Theory, Interband and Plasma Effects," *Proceedings of the Sixth International Conference on the Physics of Semiconductors, Exeter, July 16–20, 1962*, pp. 353–357 (Adlard and Son, Ltd, Dorking, 1962); also published in *Journal of Electronics and Control* **14**, (3) (1963).

P10.4 Benjamin Lax and John G. Mavroides, "Interband Magnetooptical Effects," Chapter 8 in *Physics of III-V Compounds*, Vol. 3 of *Semiconductors and Semimetals*, edited by R. K. Willardson and A. C. Beer, pp. 321–401 (Academic Press, 1967).

P10.4a C. R. Phipps Jr., S. J. Thomas, and B. Lax, *Applied Physics Letters* **25** (5), 313–314 (September 1, 1974).

P10.5 B. Lax, R.S. Eng, and N.W. Harris, "Tunable Ferrite-Loaded Electro-Optic Modulators," *IEEE Transactions on Magnetics* **27**, 5483–5485 (1991).

P10.6 Benjamin Lax, Richard S. Eng, and Neville W. Harris, "Theory of Single and Double Sideband Modulators," *Proceedings of SPIE* **1633**, Laser Radar VII: Advanced Technology for Applications, p. 206 (1992).

P10.7 Richard S. Eng, Neville W. Harris, Charles L. Summers, and Benjamin Lax, "Tunable Electro-optic Modulators for Laser Radar Applications – Experimental Results," *Proceedings of SPIE* **1633**, Laser Radar VII: Advanced Technology for Applications, p. 216 (1992).

P10.8 B. Lax, J. A. Weiss, N. W. Harris and G. F. Dionne, "Quasi-optical ferrite reflection circulator," *IEEE Transactions on Microwave Theory and Techniques* **41**, 2190–2197 (1993).

P10.9 J. A. Weiss, Neville W. Harris, Benjamin Lax, and G. F. Dionne, "Quasi-optical Reflection Circulator: Progress in Theory and Millimeter Wave Experiments," *Proceedings of SPIE* **2211**, International Conference on Millimeter and Submillimeter Waves and Applications, San Diego, January 10–11, 1994, pp. 416–427 (1994).

P10.10 Gerald F. Dionne, Gary A. Allen, Pamela R. Haddad, Caroline A. Ross, and Benjamin Lax, "Circular Polarization and Nonreciprocal Propagation in Magnetic Media," *Lincoln Laboratory Journal* **15**, 323–340 (2005).

P10.11 Benjamin Lax, "Cyclotron Resonance Spectroscopy of Semiconductors," Invited paper at the Special Session on the 50th Anniversary of Cyclotron Resonance, *Physics of Semiconductors: 27th International Conference on the Physics of Semiconductors – ICPS-27, Flagstaff, Arizona, 26–30 July, 2004*, edited by Jose Menéndez and Chris G. Van de Walle, AIP Conference Proceedings **772**, pp. 11–16 (American Institute of Physics, 2005).

Theses Supervised or Mentored by Prof. Benjamin Lax

These were done in the MIT Department of Physics, unless otherwise noted.

MIT PhD Theses Supervised by Prof. Lax

T1 Shapira, Yaacov, The influence of a magnetic field on ultrasonic dispersion in metals, 1964.

T2 Halpern, John, Interband magneto-optical effects in semiconductors, 1964.

T3 Sacks, Barry Howard, Magnetic field effects in semiconductor lasers, 1967, Electrical Engineering.

T4 Greenebaum, Michael, Magnetoplasma waves in bismuth-antimony alloys at high magnetic fields, 1967.

T5 Stimets, Richard Warren, Reflection studies of magnetoplasma-optical phonon coupling in polar semiconductors, 1969.

T6 Waldman, Jerry, Submillimeter cyclotron resonance in polar semiconductors, 1970.

T7 Reine, Marion Bartholomew, Interband magneto-reflectivity of gallium antimonide and gallium arsenide, 1970.

T8 Hart, Timothy Richard (1940–1994), Magnons and phonons in solids, 1970.

T9 Temkin, Richard J., Experimental charge density of copper, 1971.

T10 Smith, Karl Ulf Helmer, Optical properties of the lead salts, 1971.

T11 Cohn, Daniel Ross, Magneto-optical properties of polarons, 1971.

T12 Bernstein, Tsur, Effects of high magnetic fields on spin waves in antiferromagnets, 1971.

T13 Cronburg, Terry Lee, Magneto-spectroscopy in solids with far infrared lasers, 1972.

T14 Cherlow, Joel Marshall, Photoluminescence and Raman scattering in silicon, 1972.

T15 Auyang, Ying Chi Sunny, Inelastic and stimulated scattering of light from mobile carriers in semi-conductors, 1972.

T16 Lee, Neville Kar-Shek, Zeeman effect of impurities and excitons in semiconductors, 1973.

T17 Oli, Basil Akolisa, The resonance model in pseudopotential theory of transition and rare-earth metals, 1976.

T18 Yacovitch, Robert Daniel, Magnetostriction in ferromagnets and antiferromagnets, 1977.

T19 Wolfe, Stephen Mitchell, Modulated submillimeter laser interferometer system for plasma density measurements, 1977.

T20 Weiler, Margaret Horton, Magnetooptical studies of small-gap semiconductors: $Hg_{1-x}Cd_xTe$ and InSb, 1977.

T21 Hacker, Mark Paul, Gas breakdown with far infrared laser radiation in intense magnetic fields, 1977.

T22 Safran, Samuel Abraham, Theory of inelastic light scattering in magnetic semiconductors, 1978.

T23 Rosenbluh, Michael, Laser magnetic resonance in excited states of atoms and molecules, 1978.

T24 Drozdowicz, Zbigniew Marian, Two photon transitions in laser pumped submillimeter lasers, 1978.

T25 Panock, Richard Lawrence, Laser spectroscopy of Rydberg states of helium using high magnetic fields, 1979.

T26 Loter, Nicholas George, Soft X-ray diagnostics for magnetically confined CO_2 laser-heated plasmas, 1979.

T27 Karmendy, Charles Victor, Jr. (deceased), Stimulated Brillouin backscatter from laser-produced plasmas in high magnetic fields, 1980.

T28 Fleming, Mark Walter, Spectral characteristics of external-cavity-controlled semiconductor lasers, 1980.

T29 Biron, David G, High intensity laser pumping of molecular gas lasers, 1981.

T30 Le, Han Quang, Measurement of electrostatic fine structure intervals nS-nP and nP-nD of Rydberg ^{4}He, 1982.

T31 Feigenblatt, Ronald Ira, Bandgap-resonant high field magnetospectroscopy of II-VI semiconductor donors, 1982, Electrical Engineering.

T32 Danly, Bruce Gordon, Frequency tuning and efficiency enhancement of high power FIR lasers, 1983.

T33 Bezjian, Krikor, The response of metallic microgrids to incident electromagnetic radiation, 1985.

T34 Zheng, Xiao-Lu, The optical and magneto-optical properties of GaAs/(GaAl)As quantum well structures, 1989.

T35 Evangelides, Stephen George, Jr., Near resonance tuning in pulsed high power $^{12}CH_3F$ and $^{13}CH_3F$ far infrared Raman lasers, 1989.

T36 Bales, James Williams, Intersubband transition in quantum-wells, 1991.

MIT Master's Theses Supervised by Prof. Lax

T37 Galeener, Frank Lee, Longitudinal magneto-plasma reflection in semiconductors, 1962.

T38 Reine, Marion Bartholomew, Effect of magnetic and electric fields on the infrared absorption in Ge, 1965.

T39 Kurtin, Stephen Lane, High field magnetic resonance in uniaxial ferrimagnetics, 1966.

T40 Brecher, Aviva, Cyclotron resonance quantum effects in p-type semiconductors, 1968.

T41 Tichovolsky, Elihu Jerome, Oscillatory magnetoreflection from bismuth-antimony alloys, 1969.

T42 Mandel, Paul David, High resolution submillimeter-wave spectroscopy using non-collinear mixing of laser radiation, 1975.

T43 Rosenbluh, Michael, Design and evaluation of a continuous-wave, step-tuneable far infrared source for solid state spectroscopy.

T44 Alavi, Kambiz, Third-order optical susceptibility of CdTe in the 10 μm region, 1977.

T45 Ma, William Wai-Nam, CO_2 laser irradiation of solid targets in high magnetic fields, 1977.

T46 Yangos, John P., Stimulated Brillouin backscattering from a laser-produced magnetoplasma, 1977, Nuclear Engineering.

T47 Lorenzen, Col. Gary Lee, USAF (1954–2000), The behavior of $Pb_{1-x}Sn_xTe$ semiconductor laser diodes in a magnetic field, 1984, Co-Supervised by Roshan L. Aggarwal and Benjamin Lax.

T48   Mims, Verett Ann, Time-resolved optical spectroscopy of AlGaAs/GaAs heterostructures, 1988.

MIT Bachelor's Theses Supervised by Prof. Lax

T49   Stein, Richard Jay, Development of a cyanide laser, 1967, BS.
T50   Cullen, Dennis Patrick, Momentum transfer in ablated aluminum targets induced by a focused TEA CO_2 laser, 1973, BS.
T51   Schmid, Stephen Joseph, The DC magnetic field of the human head, 1977, Electrical Engineering, BS.
T52   Gomez, Camilo Ciro, Faraday rotation in the Alcator tokamaks, 1981, Electrical Engineering, BS.

MIT Theses Mentored by Prof. Lax

T53   Berk, Aristid Dimitri, Cavities and waveguides with inhomogeneous and anisotropic media, 1954, ScD, Electrical Engineering, Supervised by Richard B. Adler.
T54   Brown, Richard Neil, Measurements of effective masses of charge carriers in semiconductors by Faraday rotation, 1958, master's, Supervised by George Fred Koster.
T55   Wright, George Buford (1926–2017), Magneto-optical properties of mercury selenide, 1960, PhD, Supervised by Arthur Robert von Hippel.
T56   Halverson, Ward Dean, "A study of plasma columns in a longitudinal magnetic field," 1965, ScD, MIT Dept. of Geology, Supervised by Francis Bitter.
T57   Spitzberg, Richard Michael, On visual mechanisms and the processing of spatio-temporal information, 1973, PhD, Supervised by Whitman A. Richards.
T58   Favrot, Gervais Freret, Jr., Magnetic field reversal effect in inter-Landau level absorption and simulated spin-flip Raman scattering in n-type InSb, 1998, PhD, Supervised by Don Heiman and Peter A. Wolff.

Thesis from Another University Mentored by Prof. Lax

T59   Dexter, Richard N., The cyclotron resonance of holes in germanium and silicon, 1955, PhD, Physics, University of Wisconsin–Madison.

Notes

Chapter 1   Early Years in Miskolc, Hungary, 1915–1926

1.1   The six children in Ben's family were: Alex, Erno (Ernest), Ella, Bela (Ben), Dezso (David), and Ilona (Eleanor).

1.2   Tokaj is a town in the wine-producing area in northeastern Hungary, about 54 km east of Miskolc.

Chapter 2   School Days in Brooklyn, 1926–1936

2.1   *The Adventures of Tom Sawyer* by Mark Twain (1835–1910), published in 1876, is a classic novel about a young boy growing up along the Mississippi River.

2.2   *The Last of the Mohicans: A Narrative of 1757* is a historical novel written by James Fenimore Cooper (1789–1851) and published in 1826.

2.3   *Uncle Tom's Cabin; or, Life Among the Lowly* is a novel about slavery, written by Harriet Beecher Stowe (1811–1896) and published in 1852.

2.4   "*Tom Swift*" actually refers to a series of over 100 juvenile science fiction and adventure novels that emphasize science, invention, and technology. Written by ghostwriters, these books were packaged by the Stratemeyer Syndicate. Tom, a teenager, is the main character. By 1935, Ben would have had access to the first 38 volumes in this series, published by Grosset & Dunlap, New York starting in 1910.

2.5   *Tarzan* is actually a series of 24 adventure novels written by Edgar Rice Burroughs (1875–1950) and published by A. C. McClurg & Co., Chicago, from 1916 to 1966. By 1933, Ben would have had access to the first 16 *Tarzan* novels.

2.6   Isidor Isaac Rabi (1898–1988) is an American physicist who received the 1944 Nobel Prize in Physics for "... his resonance method for recording the magnetic properties of atomic nuclei," which is called nuclear magnetic resonance. In 1940–1945 he was Associate Director of the MIT Radiation Laboratory.

2.7   "*Erlkonig*" is a well-known German poem written by Johann Wolfgang Goethe in 1782. It is about the death of a child at the hands of a supernatural demon, the *Erlkonig*.

2.8   *Anthony Adverse*, a historical novel by William Hervey Allen, Jr. (1889–1949), published by Farrar & Rinehart in June 1933.

2.9   Samuel Stanford Manson (1919–2013) was born in Jerusalem. At age seven, he and his family immigrated to Brooklyn. A brilliant student in mathematics, he graduated Cooper-Union Institute of Technology with highest honors in 1937. In 1942 he earned an MS in mechanical engineering at the University of Michigan. His illustrious career in materials science began at the National Advisory Committee for Aeronautics (NACA) Langley Research Center in 1942, and continued in 1943 at the NACA Lewis Research Center in Cleveland.

2.10   Richard Phillips Feynman (1918–1988) was a legendary theoretical physicist who shared the 1965 Nobel Prize in Physics with Julian Seymour Schwinger (1918–1994) and Shinichiro Tomonaga (1906–1979) for "... their fundamental work in quantum electrodynamics, with deep-ploughing consequences for the physics of elementary particles." Feynman received his BS degree in physics from MIT in 1939. His BS thesis, "Forces and stresses in molecules," was supervised by John C. Slater.

2.11   Paul Malcolm Marcus (b. 1921) would go on to receive a master's degree in 1942 and a PhD in chemical physics in 1943, both from Harvard University. His PhD thesis, originally classified, was titled "Fluid motion in underwater explosions." He then worked at the MIT Radiation Laboratory from 1943 to 1946 as one of 28 members of the Theory Group (Group 43), headed by the noted physicist George E. Uhlenbeck. "Radiation Laboratory Staff Members, 1940–1945," Radiation Laboratory, MIT, Cambridge, Massachusetts; in the MIT Archives.

2.12   William Anthony Granville and Percey Franklyn Smith (both of Yale University), *Elements of the Differential and Integral Calculus* (Ginn & Co., New York, 1904).

Chapter 3   College Days—Brooklyn College and Cooper Union, 1936–1942

3.1 Harry Schwartz (1919–2004) received his BA (1940), MA (1941) and PhD (1944) from Columbia. He was a longtime staff writer and editorial writer at the *New York Times*. His focus was the Soviet Union and the Cold War. He was valedictorian of his class at Columbia. He worked for the Office of Strategic Services (OSS), the precursor to the Central Intelligence Agency. www.nytimes.com/2004/11/12/obituaries/harry-schwartz-85-times-editorial-writer-dies.html

3.2 Melvin J. Lax (1922–2002), a theoretical physicist, was Distinguished Professor of Physics at City College of New York. Before that he was at Bell Laboratories in Murray Hill, New Jersey and at Syracuse University. He received his BA in physics from New York University in 1942 (three years after Ben graduated from Cooper Union), and his MS and PhD, both in physics from MIT, in 1943 and 1947. His PhD thesis, "Production of mesons by electromagnetic means," was supervised by Herman Feshbach.

3.3 Harry Wilfred Reddick (1883–1962) and Frederic Howell Miller (1903–1964), *Advanced Mathematics for Engineers*, first edition (London, 1938).

3.4 Joaquin Mazdak Luttinger (1923–1997), a native of New York City, was an important theoretical physicist who made major contributions to condensed matter physics. He received his BS degree in 1944 and his PhD in 1947, both in physics from MIT. His PhD thesis, "Dipole interactions in crystals," was supervised by László Tisza. Ben's later interactions with Luttinger most likely were on Luttinger's contributions to the energy band structure of semiconductors, which Ben was exploring with cyclotron resonance and interband magneto-optics. The three parameters that characterize the valence band of zinc-blende semiconductors are called the Luttinger parameters.

3.5 J. M. Luttinger, born in December 1923, was actually 18 years old in the summer of 1942. The Biographical Note in his MIT PhD thesis states "In the summer of 1942, he was a student at the Brown University session in advanced mechanics."

Chapter 4   Army Days and the MIT Radiation Laboratory, 1942–1945

4.1 Philip Fox (1878–1944) was a Colonel in the Army Signal Corps and was the Commanding Officer of the Army Electronics Training Center at Harvard and MIT when Ben arrived for duty in Cambridge, Massachusetts in January 1943. Fox had extraordinarily strong academic as well as military credentials. He earned a BS in mathematics in 1897, and master's degree in 1901, both from Kansas State University. He then earned a BS in physics at Dartmouth College in 1902. He did graduate studies in astronomy at the University of Berlin and at the University of Chicago, where he received his PhD in astronomy. As an astronomer, he held several leadership positions, including Chair of the Astronomy Department at Northwestern University, and Director of the Adler Planetarium and the Chicago Museum of Science and Industry. He served in the Spanish-American War and in World War I. He retired from the Army in 1943 and continued to teach at Harvard until his untimely death in 1944 at the age of 66.

"The People who Worked in the Army Training Program at the M.I.T. Radar School, 1941–1945," in the MIT Archives. https://en.wikipedia.org/wiki/Philip_Fox_(astronomer)

4.2 Philippe Emmanuel Le Corbeiller (1891–1980) was a professor of Applied Physics at Harvard. He was an expert in the electronics of telecommunications, and in the theory and applications of nonlinear systems. He studied engineering and mathematics in France at the École Polytechnique. He came to Harvard in 1941, and taught basic electronics in the Harvard Radio Research Laboratory to Army and Navy personnel.

4.3 Ronold Wyeth Percival King (1905–2006) was a prolific professor of applied physics at Harvard who specialized in microwave antennas. Ben would have had access to two of King's classic textbooks on microwave theory and technology: *Electromagnetic Engineering* (McGraw-Hill, 1945) and, with Harry R. Mimno and Alexander H. Wing, *Transmission Lines, Antennas and Wave Guides* (McGraw-Hill, 1945).

4.4 Harry Rowe Mimno (1900–1981) received his EE degree from Rensselaer Polytechnic Institute in 1921, and his AM and PhD degrees in physics from Harvard in 1926. His PhD thesis title is "A study of the operation of vacuum tubes at relatively high power levels." He helped found and became Associate Director of the wartime Cruft Laboratory Training School at Harvard, which trained Navy personnel to be radio operators in the early to mid-1940s. He taught courses in the fields of radio propagation, radar, automatic computation, electronic navigation, and telemetry. https://ieeexplore.ieee.org/stamp/stamp.jsp?arnumber=4201646

4.5 Murray Winnick, in Ben's recollections, was an electrical engineer and an electronics expert, who was a classmate of Ben's at the Harvard Radio Research Laboratory, and who, with Ben, was assigned to the MIT Radiation Laboratory. However, Winnick does not appear in the book *Five Years at the Radiation Laboratory*, published by MIT in 1946.

4.6 William H. Radford (1909–1966) earned his BS at Drexel Institute in 1931 and his MS in electrical engineering at MIT in 1932. He was Associate Director of MIT's Radar School from 1941 to 1947. In 1951, he became a professor in the MIT Department of Electrical Engineering. He also joined Lincoln Laboratory in 1951. By 1957, he was Associate Director of Lincoln Laboratory, and in 1964 he was Director of Lincoln Laboratory. *Physics Today* **19** (7), 137 (July 1966).

4.7 MIT Radiation Laboratory operated under the supervision of the National Defense Research Committee, from October 1940 until December 31, 1945. Early developments at the Radiation Laboratory include airborne intercept radar, a gun aiming system, and aircraft navigation (LORAN). Some of the most critical contributions of the Radiation Laboratory were the microwave early-warning (MEW) radars, which effectively nullified the V-1 threat to London, and air-to-surface vessel (ASV) radars, which turned the tide on the U-boat threat to Allied shipping. It was housed in legendary wood-frame buildings on the MIT campus. At its peak in 1945, the Rad Lab employed 3,500 people and was spending close to $4 million a month. www.ll.mit.edu/about/history/mit-radiation-laboratory

4.8 John Hasbrouck Van Vleck (1899–1980) spent nearly all of his exemplary career as a professor of physics at Harvard. Together with Philip Anderson, a student of Van Vleck's, and Sir Nevill Mott, he received the 1977 Nobel Prize in Physics for "research on the electronic structure of magnetic and disordered systems."

4.9 Ernest Charles "Ernie" Pollard (1906–1997) was a distinguished professor of physics and biophysics at Yale University. From 1941 to 1945 he was at the MIT Radiation Laboratory, working on such radar projects as Li'l Abner, for which he was granted a patent. He was head of Division 10 (Ben's division), which focused on ground-based and ship-based radar development. Pollard wrote a memoir, *Radiation: One Story of the MIT Radiation Laboratory* (Woodburn Press, Durham, NC, 1982).

4.10 Jacob Millman (1911–1991), a native of Russia, earned his BS in physics from MIT in 1932 and his PhD in theoretical physics from MIT in 1935. His PhD thesis, "Electronic energy bands in metallic lithium," was supervised by John C. Slater. Millman was Professor of Electrical Engineering at Columbia, and was at the MIT Radiation Laboratory in 1942–1945. A prolific writer, he is remembered for his outstanding series of five classic textbooks on electronic circuits and electronic devices. https://ethw.org/Jacob_Millman

4.11 Louis Nicot Ridenour (1911–1959) was a physicist who pioneered the development of radar during World War II. He earned his PhD in physics from the California Institute of Technology. He was the Assistant Director of the MIT Radiation Laboratory from 1941 to 1946. In 1950 he became the first Chief Scientist of the Air Force. He was Editor in Chief of the monumental 28-volume MIT Radiation Laboratory Series, and edited Volume 1 of that series, *Radar System Engineering* (McGraw-Hill, 1947). https://en.wikipedia.org/wiki/Louis_Ridenour

4.12 The Li'l Abner Radar Set, officially designated as AN/TPS-10, was a lightweight X-band nodding height-finding radar. The transmitter used a magnetron. The display was a cathode-ray tube of 12" diameter. The MIT Radiation Laboratory developed and produced the first version of this radar near the end of World War II. Zenith produced the A-model sets in the post-war period. The large, vertically mounted antenna was three feet wide and ten feet high. Two operators were needed to run the radar set. The initial model operated at a frequency of 9,000 to 9,160 MHz and had a maximum reliable range for bombers of 60 miles at 10,000 feet. www.radartutorial.eu/19.kartei/11.ancient/karte074.en.html

4.13 Samuel W. Levine (1916–1997), a native of Dallas, Texas, received a BS in chemical engineering (1938) and an MS in chemistry (1940), both from Texas Agricultural and Mechanical University. In 1940–1946 he was a captain in the US Army Air Corps, assigned to the

Li'l Abner radar project that Ben directed at the MIT Radiation Laboratory. Coincidentally, he became Ben's brother-in-law by marrying Blossom's sister, Eleanor Cohen, on August 26, 1944. Sam earned a PhD in chemistry at MIT in 1948. He went on to an accomplished career in industry. He was Technical Director of the Fairchild Camera and Instrument Corporation, Syosset, New York in 1953–1970, when, at the direction of founder Sherman Mills Fairchild (1896–1971), he administered the Fairchild Foundation Fellowship program, which granted a full fellowship to a selected graduate student in semiconductor physics or closely related disciplines at each of eight selected institutions, including MIT, Harvard, California Institute of Technology, University of Illinois, Stanford and Purdue. Ben selected one of the editors (MBR) as the MIT recipient for the years 1966–1969.

4.14 Belmont Greenlee Farley (1920–2008) received his BS from the University of Maryland in 1941. He then began doctoral studies in mathematics at MIT, which were interrupted by the war. He joined the MIT Radiation Laboratory, developing improved radar for detecting low-flying aircraft, which he tested in England against actual low-flying Luftwaffe planes during the bombing of London. After the war, he earned both his MS (1946) and PhD (1948) at Yale University. He then joined the illustrious group headed by Shockley at Bell Telephone Laboratories that had just invented the transistor. His career continued with his later work at MIT on the Whirlwind computer, and then his groundbreaking research on neural networks and the electrophysiology of the brain. www.philly.com/philly/obituaries/20080310_Belmont_G__Farley__87__pioneer_in_computers.html

4.15 Thomas Marlett Moore received his BSME degree from Purdue in 1934. "Radiation Laboratory Staff Members, 1940–1945," Radiation Laboratory, MIT, Cambridge, MA, in the MIT Archives.

4.16 Lee Alvin DuBridge (1901–1994) was the founding director of the MIT Radiation Laboratory during 1940–1945, and was President of the California Institute of Technology in 1946–1969. He graduated Cornell College in 1922, and earned his MA (1924) and PhD (1926), both in physics, at the University of Wisconsin–Madison. His PhD thesis was "Variations in the photo-electric sensitivity of platinum." Among many advisory positions to government and the military, he was Presidential Science Advisor to Presidents Truman, Eisenhower and Nixon.

4.17 Gerald Silas Heller (1920–1992) was at the MIT Radiation Laboratory from 1942 to 1945. He received his PhD in physics from Brown University in 1948. Subsequently he taught at Brown and was at Lincoln Laboratory, where he was Group Leader of the Resonance Physics Group, during which time he and Ben would have worked together on microwave devices. He has several publications in ferrite microwave devices. https://prabook.com/web/gerald_silas.heller/229520

4.18 Herbert George Weiss received his SB degree in electrical engineering from MIT in 1940. Soon thereafter, he joined the MIT Radiation Laboratory.

4.19 Robert Vivian Pound (1919–2010) was a professor of physics at Harvard University who, in 1946, with his collaborators Edward Purcell and Henry Torrey, adapted the MIT Rad Lab microwave techniques – widely used to this day in radar and communications – to detect nuclear magnetic resonance in solids.

4.20 Albert McCavor Clogston (1917–2013), a native of Boston, earned his BS in 1938 and his PhD in 1941, both in physics at MIT. His PhD thesis, "Forced oscillations of electromagnetic cavity resonators," was supervised by Julius A. Stratton. He joined the MIT Radiation Laboratory, where he focused on magnetrons, soon becoming head of the Magnetron Research and Development Group. After the war, he went on to a distinguished research career at Bell Laboratories. www.nasonline.org/publications/biographical-memoirs/memoir-pdfs/clogston-albert.pdf

4.21 Melvin Arnold Herlin (1923–2013), a 1943 graduate of the University of Utah, received his PhD in physics at MIT in 1948. His PhD thesis, "Breakdown of a gas at microwave frequencies," was supervised by Sanborn C. Brown. He was at the MIT Radiation Laboratory during 1943–1945. He was on the faculty of the MIT Physics Department during 1945–1955, and then at Lincoln Laboratory. He worked in the field of radar research for 44 years.

4.22 Edward Mills Purcell (1912–1997) was a legendary physicist. With MA and PhD degrees in physics from Harvard, during World War II he worked at the MIT Radiation Laboratory on the development of microwave radar, after which he returned to Harvard for a brilliant academic career. He shared the 1952 Nobel Prize for Physics for his discovery of nuclear magnetic resonance in liquids and in solids.

4.23 Project Cadillac, the highest priority program at the MIT Radiation Laboratory, was developing an airborne radar set that would provide longer range detection of aircraft. It was named for Cadillac Mountain in Maine, where the sun first rises on the United States.

4.24 Richard M. Walker received his BS in electrical engineering in 1943 from the University of Kansas. Ben met him at the MIT Radiation Laboratory. In 1950 Walker cofounded Microwave Associates. Headquartered in Lowell, Massachusetts, the company, now called MACOM, today is a major developer and manufacturer of RF, microwave and millimeter wave semiconductors, components and technologies. It has 1,100 employees.

Chapter 5   Graduate School in Physics at MIT, 1946–1949

5.1 John William Marchetti (1908–2003), born in Boston to immigrant parents, had distinguished careers both in the military and in industry. He completed a six-year combined undergraduate and graduate program at Columbia College, receiving undergraduate degrees and an Electrical Engineering degree in 1931. His technical expertise was radar, and he led several successful developments of prototype radar systems that rapidly achieved operational status during the war years, including efforts at the MIT Radiation Laboratory. In 1945, Major Marchetti was assigned as Acting Commanding Officer of the newly formed Cambridge Field Station. He would soon become the founding Technical Director of the Air Force Cambridge Research Laboratories. https://en.wikipedia.org/wiki/John_W._Marchetti   www.college.columbia.edu/cct_archive/sep03/obituaries.html

5.2 In September 1945, the Army Air Corps established a new organization, named the Cambridge Field Station, to support the Army Signal Corps with continued research on radio, radar, and electronics. By June 1946 it was staffed with 350 of the many scientific, engineering, and technical support personnel who had been at the MIT Radiation Laboratory and the Harvard Radio Research Laboratory during the war. It was organized into two Directorates, Electronics Research and Geophysics Research. This new organization was located in buildings at 224 and 230 Albany Street, at the edge of the MIT campus. Coincidentally, the National Magnet Laboratory, of which Lax would be the founding director, would be located at 166 Albany Street, just a few doors away. In July 1949, the Cambridge Field Station was renamed the Air Force Cambridge Research Laboratories, and in the early 1950s migrated to its permanent location at Hanscom Air Force Base in Bedford, MA, where it became one of the world's finest research laboratories. Edward E. Altshuler, *The Rise and Fall of Air Force Cambridge Research Laboratories* (CreateSpace Independent Publishing Platform, 2013).

5.3 Albert G. Hill (1910–1996) was Professor Emeritus in the MIT Physics Department. He was a leader in the development of radar in World War II. He was Director of Lincoln Laboratory during 1952–1955. http://news.mit.edu/1996/hill-1030

5.4 Julius Adams Stratton (1901–1994) earned his SB in 1923 and SM in 1926, both in electrical engineering at MIT. He received his ScD in mathematical physics in 1928 from Eidgenossische Technische Hochschule, Zurich, Switzerland. He was the 11th president of MIT, 1957–1966, during which period the National Magnet Laboratory was founded. His classic textbook, *Electromagnetic Theory*, was first published in 1941. https://libraries.mit.edu/mithistory/institute/offices/office-of-the-mit-president/julius-adams-stratton-1901–1994/ www.nap.edu/read/12562/chapter/18

5.5 John Clarke Slater (1900–1976) received a BS in physics from the University of Rochester in 1920 and a PhD in physics from Harvard in 1923. A legendary theoretical physicist who focused on the electronic structure of solids, he was recruited from Harvard to head the Physics Department at MIT in 1930, where he assembled a prodigious group of physicists who would set a new standard for undergraduate education in physics. He wrote 14 books, including three early seminal textbooks with his MIT colleague Prof. Nathaniel H. Frank (1903–1984). https://en.wikipedia.org/wiki/John_C._Slater www.nasonline.org/publications/biographical-memoirs/memoir-pdfs/slater-john.pdf

5.6 Louis Dijour Smullin (1916–2009) received a BS degree from the University of Michigan, Ann Arbor, in 1936, and an SM degree from MIT in 1939, both in electrical engineering. His MIT master's thesis, "The acceleration and focusing of electrons in multi-stage tubes," was supervised by Prof. John G. Trump. He spent most of his career at MIT, including the MIT Radiation Laboratory. He assisted in the formation of the MIT Lincoln Laboratory. He headed the MIT Department of Electrical Engineering and

Computer Science from 1966 to 1974. https://en.wikipedia.org/wiki/Louis_Smullin https://eecs-newsletter.mit.edu/articles/2009-fall/remembering-louis-dijour-smullin-1916–2009/

5.7 Malcom Woodrow Pershing Strandberg (1919–2015) received his BS degree from Harvard College in 1941 and his PhD degree in physics from MIT in 1948. He spent his entire career at MIT, beginning in 1941 at the MIT Radiation Laboratory, where he developed microwave radar, and continuing in the Department of Physics after 1945. His PhD thesis, "Research on resonant molecular absorption in the microwave region," was supervised by Prof. A. G. Hill. His career focused on microwave spectroscopy and solid state physics, and more recently on the physics of biological systems. http://web.mit.edu/physics/people/faculty/strandberg_malcom.html

5.8 Sanborn "Sandy" Conner Brown (1913–1981), a 1937 graduate of Dartmouth College, received his PhD in physics from MIT in 1944. His PhD thesis, "A theory of the initial avalanche in the breakdown of a discharge counter in helium," was supervised by Prof. Robley D. Evans. His career was spent in the MIT Physics Department, where he built up a large and productive research group focused on the physics of ionized gases. The MIT Barton library lists a total of 94 theses supervised by Prof. Brown between 1942 and 1971, including 34 doctoral theses. www.nytimes.com/1981/12/02/obituaries/dr-sanborn-c-brown-expert-on-ionized-gas.html

5.9 William Phelps Allis (1901–1999) was a theoretical physicist who specialized in the electrical discharge of ionized gases. He was a professor of physics at MIT for most of his career. He influenced Ben's thesis research. See: *Electrons, Ions, and Waves: Selected Works of William Phelps Allis*, Sanborn C. Brown, Editor (MIT Press, 1967).

5.10 John Mark Deutch (1938-) was born in Brussels, Belgium. He became a US citizen in 1945. He received a BS degree in history and economics from Amherst College. He earned a BS degree in chemical engineering in 1961 and a PhD in chemistry in 1966, both from MIT. His PhD thesis, "Selected problems in statistical mechanics," was supervised by Irwin Oppenheim. His career at MIT was mainly in administration; he serving as Provost and Dean of Science. He had a number of positions in the Department of Energy in Washington, DC. In the Clinton Administration he was Deputy Secretary of Defense from 1994 to 1995, and Director of Central Intelligence Agency from 1995 until 1996. https://en.wikipedia.org/wiki/John_M._Deutch

5.11 Hans Mueller (1900–1965) was an optical physicist and professor at MIT. He joined the Physics Faculty at MIT in 1925, and became a popular teacher. https://en.wikipedia.org/wiki/Hans_Mueller_ (physicist)

5.12 Bertram Eugene Warren (1902–1991), a native of Waltham, Massachusetts, was a distinguished physicist and brilliant X-ray crystallographer who spent his 40-year career as professor of physics at MIT. He is comparable to others in his field, including Ewald, Laue, Bragg, Debye, and Pauling. He was revered for both his teaching as well as his research. He received his BS in 1925 and his ScD in 1929, both in physics and both at MIT. His ScD thesis, "An X-ray determination of the structure of the metasilicates," was mentored by Sir William Lawrence Bragg, CH, OBE, MC, FRS (1890–1971) while Bragg was visiting at MIT. Leonard Muldawer, "Bertram Eugene Warren, 1902–1991," *Journal of Applied Crystallography* **29**, 309–310 (1996). https://doi.org/10.1107/S0021889896003329 https://journals.iucr.org/j/issues/1996/04/00/

5.13 Robley Dunglison Evans (1907–1995) was a nuclear physicist who received his BS (1928), MS (1929) and PhD (1932) degrees, all in physics and all from the California Institute of Technology. He spent his entire career in the Physics Department at MIT, mainly doing research on the effects of radiation and radioactive materials on the human body. He is the author of the classic textbook, *The Atomic Nucleus* (McGraw-Hill, 1955). Lax recalls studying Evans' book in early 1947, so perhaps he had an early version of the book.

5.14 Nathaniel Herman Frank (1903–1986), a native of Boston, earned his BS in electrochemical engineering at MIT in 1923, and his ScD in physics at MIT in 1927. He spent his entire career in physics at MIT, including a few years at the MIT Radiation Laboratory. His research interests were theoretical physics and metallic conduction, but he was particularly interested in physics education. In the 1950s, he was a member of the Physical Science Study Committee at MIT, which revitalized high school physics curricula. He is credited with a marvelous teaching ability. He was a coauthor with his MIT colleague John Slater of three seminal physics textbooks. In his obituary for Frank (*Physics Today* **37** (10), 112 (October 1984), his MIT colleague Philip Morse notes that one of the standard questions Frank would always ask was, "What does your answer mean physically?"

5.15 The PhD qualifying exam in the Physics Department at MIT consisted of two parts. First came the written exam, covering selected areas of physics and usually requiring months of preparation, including working problems from previous written exams. Then came the oral exam, typically three hours, in front of a three-member committee headed and assembled by one's thesis advisor. At the end, the student waits outside the room while the committee caucuses, and then is brought back in to receive the verdict.

5.16 Philip McCord Morse (1903–1985), a native of Shreveport, Louisiana, was a legendary physicist who was a professor of physics at MIT for 30 years. His many roles in advising and participating in government included directing the Office of Operations Research during World War II. Referred to as a generalist, he made major contributions to a number of fields of physics, including acoustics and the theory of sound absorption, astrophysics, quantum mechanics, mathematical tables, and atomic power. He was a member of the National Academy of Sciences. Among the many influential books he wrote is the classic volume *Methods of Theoretical Physics* (McGraw-Hill, 1953), coauthored with Herman Feshbach (1917–2000), also a professor of physics at MIT. https://en.wikipedia.org/wiki/Philip_ M._Morse

5.17 George G. Harvey (1908–1988) joined the MIT Physics Faculty in 1934. He had a long affiliation with the MIT Research Laboratory of Electronics, ultimately serving as Associate Director. He was well known for his research on X-ray scattering and atomic structure as well on electron microscopy. He supervised four PhD theses and three master's theses between 1952–1961, mostly in the optical studies of metals.

5.18 Arthur C. Hardy (1895–1977) was a professor of physics at MIT for 44 years, and was Chair of the MIT Physics Department in the 1930s. He was a specialist in optics and photography. He developed the spectrophotometer in 1927. He helped develop sound recording on film for motion pictures. With Fred H. Perrin, he wrote the classic textbook *The Principles of Optics* (McGraw-Hill, 1932). www.osa.org/ en-us/history/biographies/arthur-c--hardy/

5.19 Francis Weston Sears (1898–1975) was a professor of physics at MIT for 35 years. He was regarded by all as an outstanding physics teacher. He is remembered for his wonderful series of classic physics undergraduate textbooks, nearly all published by Addison-Wesley, the most noted of which is *University Physics*, first published in 1949 and cowritten with Mark Waldo Zemansky (1900–1981), a longtime professor of physics at the City College of New York. This classic textbook is often referred to simply as "Sears and Zemansky." In 1947, Lax would have had access to Sears' books on optics and on electricity and magnetism. *Physics Today* **29** (2), 65 (February 1976). https://en.wikipedia.org/wiki/Francis_Sears

5.20 James Dillon Cobine, *Gaseous Conductors – Theory and Engineering Applications* (McGraw-Hill, 1941).

5.21 Alexander Daniel MacDonald (1923–2005), a native of Nova Scotia, received his BS from Dalhousie University in Halifax, and his PhD in physics from MIT. He was a professor of physics and math at Dalhousie. He was Chief Scientist at the Lockheed Palo Alto Research Center, and a member of the Home Brew Computer Club.

5.22 Alexander Daniel MacDonald, "High Frequency Ionization Coefficients in Gases," PhD Thesis, MIT Department of Physics, 1949, supervised by Sanborn C. Brown.

5.23 A. D. MacDonald and S. C. Brown, "High Frequency Gas Discharge Breakdown in Helium," *Physical Review* **75**, 411–418 (February 1, 1949); A. D. MacDonald and S. C. Brown, "High Frequency Gas Discharge Breakdown in Hydrogen," *Physical Review* **76**, 1634–1639 (December 1, 1949).

5.24 Paschen's law (or Paschen's curve) is an equation that gives the breakdown voltage (the voltage necessary to start a discharge or electric arc) between two electrodes in a gas as a function of pressure and gap length. It is named after Louis Carl Heinrich Friedrich Paschen (1865–1947) who discovered it empirically in 1889. https://en.wikipedia.org/wiki/Paschen%27s_law

5.25 Henry Margenau (1901–1997) was a professor of physics at Yale with research accomplishments in metallurgy, plasmas and solid state physics. He wrote a classic graduate-level textbook with his Yale colleague, George Moseley Murphy (1905–1968), *The Mathematics of Physics and Chemistry* (Van Nostrand, 1943). Lax would have had access to this well-regarded book in his graduate student years.

Chapter 6   Postdoctoral Work at Air Force Cambridge Research Laboratories, 1949–1951

6.1 Osmund "Oz" Theodore Fundingsland (1915–1988), a native of South Dakota, worked at the MIT Radiation Laboratory during World War II. He was a contemporary MIT physics graduate student of Lax's in Sanborn Brown's group. He received his master's in 1950. His thesis was "Measurements of the collision cross-sections for electrons and gas molecules at thermal energies." His career in industrial physics unfolded at Sylvania and EG&G. In 1972, he joined the Government Accounting Office in Washington, DC, where he eventually became Chief Science Advisor to one of the GAO's divisions and a widely respected authority on the federal science budget. *Physics Today* **41** (3), 118 (March 1988).

6.2 David Atlas (1924–2015) was a widely accomplished meteorologist and one of the pioneers of the application of radar to meteorology. During World War II he was in the US Army Air Corps and worked on radar development. After the war he remained with the Air Force, working at the Air Force Cambridge Research Laboratories on radar for weather applications.

6.3 George E. Austin (1921–2011), a native of Boston, Massachusetts, was a nuclear physicist who worked on radar during World War II. He was the father of Tracey Austin, noted professional tennis player.

6.4 Albert Simon, "Ambipolar Diffusion in a Magnetic Field," *Physical Review* **98**, 317–318 (1955).

6.5 The Aerobee rocket was an unguided two-stage rocket, 18 feet in length, used for high atmospheric and cosmic radiation research. It was funded by the US Navy. The first Aerobee rocket was launched in November 1947.

6.5a It is highly probable that this is Prof. Leon Harold Fisher (1918–2012), who received his PhD in physics from University of California at Berkeley in 1943. He was a professor of physics at New York University for the years 1946-1961, working in the physics of gas discharges. He founded the successful annual Gaseous Electronics Conference, first held October 27–29, 1948 at Brookhaven National Laboratory, Upton, New York. Lax attended this first conference, and he presented a paper at the second such conference held November 3–5, 1949 in Pittsburgh, Pennsylvania.

6.6 Donald Howard Menzel (1901–1976) was a well-known theoretical astronomer. He earned a PhD in astrophysics at Princeton University in 1923. He taught at Harvard for most of his career. He is the author of the popular book on astronomy, *A Field Guide to the Stars and Planets* (HarperCollins, 1975). https://en.wikipedia.org/wiki/Donald_Howard_Menzel

6.7 David Joseph Bohm FRS (1917–1992) was among the leading theoretical physicists. He was a member of the theoretical physics group directed by Robert Oppenheimer at the University of California,

Berkeley, where he earned his PhD in 1943. Among his many accomplishments was establishing the foundation for plasma theory. https://en.wikipedia.org/wiki/David_Bohm

6.8 Eugene P. Gross (1926–1991) was a well-known theoretical physicist. He earned his PhD at Princeton in 1948, where he was one of the first graduate students of David J. Bohm. Gross spent most of his career at Brandeis University. His research focused on quantum liquids, plasmas, solids, liquid helium, and the kinetic theory of gases. In 1948–1949 he was a Carnegie Fellow at Harvard, and in 1950–1951 a Research Associate at MIT, which is when Gross attended one of Lax's lectures. https://en.wikipedia.org/wiki/Eugene_P._Gross

6.9 The Vlasov equation is a time-dependent differential for the distribution function of a plasma that consists of charged particles with a long-range Coulomb interaction. It is named after the Russian theoretical physicist Anatoly Alexandrovich Vlasov (1908–1975) who first suggested it in 1938. https://en.wikipedia.org/wiki/Vlasov_equation

6.10 D. Bohm and E. P. Gross, "Plasma Oscillations as a Cause of Acceleration of Cosmic-Ray Particles," *Physical Review* **74**, 634 (1948); D. Bohm and E. P. Gross, "Theory of Plasma Oscillations. B. Excitation and Damping of Oscillations," *Physical Review* **75**, 1864 (1949).

6.11 Ira Borah Bernstein (1924–) is a well-known theoretical plasma physicist. He received his PhD from New York University in 1950. From 1950 to 1954, when Lax would have consulted him, he was working at the Westinghouse Research Laboratories in Pittsburgh.

6.12 Theodore David Holstein (1915–1985) was a well-known condensed matter physicist. He received his PhD in physics from New York University in 1940. He was a member of the National Academy of Sciences and the American Academy of Arts and Sciences. In the early 1950s, when Lax recalls consulting him about the idea of plasma confinement with magnetic fields, Holstein would have been at the Westinghouse Research Laboratories in Pittsburgh.

Chapter 7 MIT Lincoln Laboratory, 1951–1965

7.1 MIT Lincoln Laboratory was formed in 1951 as a Federally Funded Research and Development Center, created and managed by MIT at the request of the Air Force. Its focus was on improving the nation's air defense systems through advanced electronics. It has made major contributions in science and technology in a broad range of disciplines, including radar, aerospace technology,

communications, optics, materials, computer hardware and soft-
ware, lasers, and solid state physics. Ben joined the solid state
physics group at Lincoln Laboratory in 1951, shortly after it was
created, when it was temporarily housed in buildings on the MIT
main campus in Cambridge, Massachusetts.

7.2 Jerome Bert Wiesner (1915–1994) was a Professor of Electrical
Engineering at MIT. In 1946 to 1961 he had various leadership positions
in the MIT Research Laboratory of Electronics. He would become
Chairman of President Kennedy's Science Advisory Committee in
1961–1964, and would become the 13th president of MIT (1971–1980).

7.3 Richard Brooks Adler (1922–1990) was Distinguished Professor in the
MIT Department of Electrical Engineering and Computer Science. He
taught at MIT from 1949 until his untimely death in 1990. In the 1960s,
Prof. Adler established and directed the Semiconductor Electronics
Education Committee, which produced a series of milestone textbooks
that first brought transistor-based solid state electronics into the main-
stream undergraduate engineering curriculum. Prof. Adler was Leader
of the Solid State and Transistor Group at MIT Lincoln Laboratory
from 1951 to 1953. It was Prof. Adler's group that Ben Lax joined as a
new hire in 1951. http://tech.mit.edu/V110/N2/adler.02n.html

7.4 Clarence Lester "Les" Hogan (1920–2008) was a well-known and
brilliant physicist, and was a pioneer in both microwave and semiconductor technology. Ben says that Hogan visited MIT to give a lecture
on ferrites around 1951. Hogan had just completed his PhD at Lehigh
University in 1950, and would have been working in 1951 at Bell Labs
and inventing the microwave gyrator. Shortly thereafter, in 1953, he was
named the Gordon McKay Professor in Applied Physics at Harvard. In
1958, he joined Motorola as general manager of their semiconductor
operation. www.nytimes.com/2008/08/16/technology/16hogan.html

7.5 J. O. Artman and P. E. Tannenwald were staff scientists at Lincoln
Laboratory who were working in ferrites and who collaborated with
Ben. In their 1955 paper, they give credit to Ben's calculations: J.
O. Artman and P. E. Tannenwald, "Measurement of Susceptibility
Tensor in Ferrites," *Journal of Applied Physics* **26**, 1124 (1955).

7.6 William Bradford Shockley, Jr. (1910–1989) was a brilliant solid
state physicist who led his group at Bell Telephone Laboratories to
invent the transistor. Their invention replaced the vacuum tube and
revolutionized electronics. Shockley shared the 1956 Nobel Prize in
Physics with his co-inventors John Bardeen (1908–1991) and Walter
H. Brattain (1902–1987). Shockley received his PhD in physics from

MIT in 1936. His thesis, "Electronic bands in sodium chloride," was supervised by John C. Slater. Shockley's book, *Electrons and Holes in Semiconductors, with Application to Transistor Electronics* (Van Nostrand, 1950), published the year before Ben joined Lincoln Laboratory, soon became the bible for the rapidly expanding field of semiconductor materials and device research. www.nobelprize.org/prizes/physics/1956/shockley/biographical/

7.7 Jerrold Reinach Zacharias (1905–1986) was an Institute Professor of Physics at MIT. In 1956, he inaugurated the Physical Science Study Committee, which dramatically altered the teaching and education of physicists.

7.8 Kenneth J. Button (1922–2010) was a solid state and plasma physicist. He joined Lincoln Laboratory in 1951, the same year as Ben. He was a close colleague of Ben's throughout their parallel careers at Lincoln Laboratory and the National Magnet Laboratory. They authored the classic book *Microwave Ferrites and Ferromagnetics* (McGraw-Hill, 1962). https://en.wikipedia.org/wiki/Kenneth_Button_(physicist)

7.9 Karl Lark-Horovitz (1892–1958) was head of the Physics Department at Purdue University from 1928 until 1958. He established the Physics Department at Purdue as an international leader in semiconductor materials and device research. His group is particularly noted for its pioneering work on germanium diodes for microwave detectors. www.encyclopedia.com/science/dictionaries-thesauruses-pictures-and-press-releases/lark-horovitz-karl

7.10 Hsu-Yun Fan (1912–2000) was a distinguished member of the Physics Department at Purdue University and author of the well-known textbook *Elements of Solid State Physics* (Wiley, 1987). He received his MS degree in 1934 and his ScD degree in 1937, both from MIT and both in electrical engineering. His ScD thesis was "The transition from glow discharge to arc," was co-supervised by R. D. Bennett, A. von Hippel, and W. B. Nottingham. *Physics Today* **54** (10), 90 (October 2001).

7.11 Siegfried F. Neustadter (1923–2012) received his BS degree in mathematics, physics and chemistry and his PhD in mathematics, both from the University of California, Berkeley. He was a staff member at Lincoln Laboratory who worked with Ben on the transient response of a p-n junction diode.

7.12 Herbert J. Zeiger (1925–2011) joined Lincoln Laboratory in 1953 after receiving his PhD in physics at Columbia University in the group headed by Professor Charles H. Townes. Herb remained at

Lincoln Laboratory until his retirement in 1990. https://translate. google.com/translate?hl=en&sl=de&u=https://de.wikipedia.org/ wiki/Herbert_Zeiger&prev=search

7.13 Simon "Si" Foner (1925–2007) was an experimental solid state physicist who worked in magnetism and superconductivity. Si joined Lincoln Laboratory as a staff physicist in 1953. In 1961 he became a founding staff physicist at the National Magnet Laboratory. His obituary in *Physics Today* describes him as a "tireless tinkerer with experimental apparatus, restless gadfly, uncontrollable (and unredeemable) punster." Si could almost always be seen at the Magnet Laboratory wearing a bow tie and white shirt with the sleeves partially rolled up. He seemed to be in a pleasant state of constant motion. https://physicstoday.scitation.org/do/10.1063/PT.4.2146/full/

7.14 Robert H. Fox was a member of Ben's group on ferrites at Lincoln Laboratory.

7.15 James W. Meyer was a member of Ben's group on ferrites at Lincoln Laboratory.

7.16 Richard N. Dexter is a Professor Emeritus in the Physics Department at the University of Wisconsin-Madison. He received his PhD in physics from the University of Wisconsin-Madison in 1955, although his actual thesis work was done at Lincoln Laboratory under the direction of Ben Lax: Richard N. Dexter, "The cyclotron resonance of holes in germanium and silicon," PhD Thesis, University of Wisconsin-Madison (1955).

7.17 W. Shockley, "Cyclotron Resonances, Magnetoresistance, and Brillouin Zones in Semiconductors," *Physical Review* **90**, 491 (May 1, 1953), Letter.

7.18 Charles Kittel (1916–2019) has had a major influence on solid state physics, both through his graduate students at the University of California, Berkeley and through his family of excellent textbooks, perhaps most notably his *Introduction to Solid State Physics*, first published in 1953 and now still in print in its eighth edition. Kittel began his research career at Bell Telephone Laboratories, but moved to UC–Berkeley in 1951 where he formed his legendary group.

7.19 It was only a very few months after Kittel visited Lincoln Laboratory in the summer of 1953 that his group at Berkeley published the first paper on the experimental observation of cyclotron resonance in a semiconductor. They used germanium samples at liquid helium temperature. Their Letter to the Editor was published in the November 1,

1953 issue of *Physical Review*: G. Dresselhaus, A. F. Kip, and C. Kittel, "Observation of Cyclotron Resonance in Germanium Crystals," *Physical Review* **92**, 827 (November 1, 1953). The manuscript was received September 8, 1953.

7.19a "A prime objective of the Berkeley solid-state physics group (consisting of Arthur Kip and myself) from 1951 to 1953 was to observe and understand cyclotron resonance in semiconductors." Charles Kittel, "Cyclotron Resonance and Structure of Conduction and Valence Band Edges in Silicon and Germanium," pp. 563–565, contained in "Appendix A: Pioneers of Semiconductor Physics Remember," in: Peter Yu and Manuel Cardona, *Fundamentals of Semiconductors: Physics and Materials Properties*, 4th edition (Springer, 2010).

7.20 Harvey Brooks (1915–2004) was the Gordon McKay Professor of Applied Physics at Harvard University during 1950–1957, when Ben would have consulted with him about pursuing cyclotron resonance. Brooks made major contributions to the theory of semiconductors.

7.21 Jacob Earl Thomas, Jr. (1918–2011) was a professor in the MIT Electrical Engineering Department in the late 1940s. In 1951–1952 he worked for the Bell Telephone Laboratories in New Jersey where he helped to develop the bipolar junction transistor. In 1952 he came to Lincoln Laboratory, continuing his work on transistors until 1955. www.legacy.com/obituaries/theithacajournal/obituary.aspx?n=jacob-earl-thomas&pid=154615369&fhid=7263

7.22 The historical origins of cyclotron resonance are described by Paul Hoch in his chapter, "The Development of the Band Theory of Solids, 1933–1960," Chapter 3 of *Out of the Crystal Maze: Chapters from The History of Solid State Physics*, edited by Lillian Hoddeson, Ernst Braun, Jurgen Teichmann, and Spencer Weart, pp. 182–235 (Oxford University Press, 1992). The idea of cyclotron resonance, that is, the resonant absorption by a system of electrons between energy levels quantized by an external magnetic field, was mentioned in a 1951 paper by the Russian physicist Jakov Dorfman and in the 1951 PhD thesis of Robert Balson Dingle (1926–2010) at Cambridge University, but was generally regarded as too difficult to be practical or even possible. Experiments attempted at Bell Telephone Laboratories by Harry Suhl and Gerald L. Pearson failed, most probably because their sample temperature, 77 K, was too high. Hoch credits the Berkeley team with the first publication [7.19] on cyclotron resonance and the Lincoln Laboratory team with the second paper [P7.4], but one that was more complete and more thorough because it reported the first data for the anisotropy of the electron effective mass.

7.23 Elias "Eli" Burstein (1917–2017) was a highly influential semiconductor physicist. Between 1943 and 1953 he headed the semiconductor research division at the US Naval Research Laboratory in Washington, DC. There his group did early research on cyclotron resonance and interband magneto-optics in semiconductors such as germanium and silicon, and was a friendly yet strong competitor to Ben Lax's group at Lincoln Laboratory. In 1953 he joined the University of Pennsylvania faculty as a professor of physics, from which he retired in 1988. The Burstein-Moss shift, in which heavy doping causes the optical absorption edge to shift to higher energy, is named after him. *Physics Today* **71** (1), 62 (January 2018). www.nytimes.com/2017/06/25/science/elias-burstein-dies-physicist.html

7.24 Charles Kittel (1916–2019), "The Solid-State Cyclotron," Invited Paper at the Annual Meeting of the American Physical Society, held at Columbia University, New York City, January 28–30, 1954; *Physical Review* **94**, 768 (May 1, 1954).

7.25 Homer F. Priest was a research chemist and materials scientist at Lincoln Laboratory. He published on a wide variety of metallurgical problems, including contamination in germanium and silicon crystals.

7.26 G. Dresselhaus, A. F. Kip, and C. Kittel, "Spin-Orbit Interaction and the Effective Masses of Holes in Germanium," *Physical Review* **95**, 568–569 (July 15, 1954), Received May 21, 1954. Letter.

7.27 The Third International Conference on the Physics of Semiconductors, ICPS-3, was held in Amsterdam, June 29–July 3, 1954. The proceedings were published in book form by the Netherlands Physical Society in 1954 and as a special issue, *Physica*, **20** (11), 801–1138 (December 1954).

7.28 Frank Herman, "Some Recent Developments in the Calculation of Crystal Energy Bands – New Results for the Germanium Crystal," Proceedings of the Third International Conference on the Physics of Semiconductors, Amsterdam, June 29–July 3, 1954; *Physica* **20** (11), 801–812 (December 1954).

7.29 G. Dresselhaus, A. F. Kip, C. Kittel, and G. Wagoner, "Cyclotron and Spin Resonance in Indium Antimonide," *Physical Review* **98**, 556–557 (April 15, 1955), received February 7, 1955, Letter.

7.30 Going from microwave frequencies to the infrared has two major benefits for cyclotron resonance. It is easier to get good resolution, especially with samples having poor lifetime. And is easier to avoid the onset of plasma formation, thereby allowing the use of more heavily doped samples.

7.31 Peter Kapitza (1894–1984) was a Russian physicist who received the Nobel Prize in Physics in 1978 for his work in low temperature phenomena. While at the Cavendish Laboratory at the University of Cambridge in the 1920s, he pioneered the creation of strong magnetic fields by applying short pulses of high current into air-core electromagnets.

7.32 John Kirtland Galt (1920–2003), PhD-Physics, MIT, 1947. Galt and his colleagues at Bell Telephone Laboratories published papers on cyclotron resonance. For example, J. K. Galt, W. A. Yager, and H. W. Dail, Jr., "Cyclotron Resonance Effects in Graphite," *Physical Review* **103**, 1586–1587 (September 1, 1956), Letter.

7.33 Laura Maurer Roth received her PhD in physics from Radcliffe College in 1957. Her thesis title is "The scattering of conduction electrons by impurities in metals." She joined Ben Lax's group at Lincoln Laboratory in 1957, but seems to have worked with Ben in prior summers, as early as 1954. She would later join the physics faculty of Tufts University, and then SUNY at Albany, where she is now Professor Emeritus of Physics.

7.34 Solomon Zwerdling (1922–2003) was a staff member at Lincoln Laboratory who worked with Ben Lax on the early interband magneto-optical experiments. He was author or coauthor with Ben on 17 publications between 1956 and 1960. In 1964 he left Lincoln Laboratory for the MIT Center for Materials Science and Engineering. Later, he joined the Jet Propulsion Laboratory in Pasadena, California.

7.35 Robert J. Keyes (1927–2012) joined MIT Lincoln Laboratory in 1950. He specialized in optical and solid state physics. He is perhaps best known for his work on the light emitting diode (LED). Keyes, along with Lax, is a coauthor of the groundbreaking paper from the Lincoln Laboratory group that reported their first GaAs LED [P.27]. Keyes edited the highly regarded book *Optical and Infrared Detectors* in the series *Topics in Applied Physics*, 19 (Springer-Verlag, 1980). He retired in 1992 as Senior Scientist. https://physicstoday.scitation.org/do/10.1063/PT.5.6060/full/

7.36 Zinc is an acceptor impurity in germanium with an ionization energy of 0.03 eV. Zinc-doped germanium makes a photoconductive infrared detector with a response extending out to a wavelength of about 35 μm. Operation at liquid helium temperature is necessary to achieve useful signal-to-noise ratios.

7.37 Henry Herbert Kolm (1924–2010) was an Austrian-born physicist who spent his entire career at MIT. His family immigrated to the US in 1939. His work on generation of high magnetic fields enabled Lax's early experiments at Lincoln Laboratory in cyclotron resonance and interband magneto-optics of semiconductors and semimetals. Kolm co-founded and was a Senior Scientist at the National Magnet Laboratory. One of the editors (DTS) recalls: "Henry and I were good friends. He was an exceptionally gifted researcher. He had many new ideas about new fields of applied physics: magnetic separation, magnetically levitated trains, electromagnetic missile accelerators, etc. However, his bad feelings about Ben (as voiced, for example, on Henry's web site) seemed to me to be unwarranted. It seemed to me that Ben supported his effort. Henry sometimes was able to get his own grants for various projects, but when he didn't Ben supported him from the main NML grant." https://henrykolm.weebly.com/ www.legacy.com/obituaries/WickedLocal-Sudbury/obituary.aspx?p age=lifestory&pid=144598916

7.38 Francis Bitter (1902–1967) had a long and varied career at MIT, beginning in 1934. He was a professor in multiple departments, including metallurgy, physics, and geophysics. He set up the original magnet laboratory in the basement of Building 4 on the MIT campus, where he devised and implemented a new design for a water-cooled electromagnet for generating high magnetic fields. That innovative and successful design has come to be called the Bitter magnet. This magnet and his magnet laboratory were the springboards for the National Magnet Laboratory that Ben Lax and Prof. Bitter collaborated on and which opened in 1963 as a national facility. That laboratory was renamed the Francis Bitter National Magnet Laboratory after Prof. Bitter died in 1967.

See: Thomas Erber, "Francis Bitter: A Biographical Sketch," pp. 3–19 in *Francis Bitter: Selected Papers and Commentaries*, edited by Thomas Erber and Clarence M. Fowler (MIT Press, 1969). https://en.wikipedia.org/wiki/Francis_Bitter

7.39 Simon Foner and Henry H. Kolm, "Coil for Pulsed Megagauss Fields," *Review of Scientific Instruments* **27**, 547 (1956); Simon Foner and Henry H. Kolm, "Coils for the Production of High-Intensity Pulsed Magnetic Fields," *Review of Scientific Instruments* **28**, 799 (1957).

7.40 Electron-hole droplets were reported in 1966 by J. R. Haynes of Bell Laboratories. Electron-hole droplets are a condensed phase of excitons in semiconductors. J. R. Haynes, "Experimental Observation of the Excitonic Molecule," *Physical Review Letters* 17 (16), 860–862 (October 17, 1966). Haynes collaborated with Shockley on what is now called the Haynes-Shockley effect.

7.41 Solomon J. Buchsbaum (1929–1993) was a distinguished physicist who chaired the White House Science Council under Presidents Ronald Reagan and George H. W. Bush. He was a senior executive at Bell Laboratories. His area of physics was gaseous and solid state plasmas. He obtained his PhD in physics from MIT in 1957. He then spent a year at the MIT Research Laboratory of Electronics and was at Bell Laboratories after that, which is when Lax most likely would have encountered him. His PhD thesis, "Interaction of electromagnetic radiation with a high density plasma," was supervised by Prof. Sanborn C. Brown, who also was Ben's thesis advisor. In 1973 he was named to the National Academy of Engineering, and in 1974 to the National Academy of Sciences. https://history.aip.org/phn/11501001.html www.nasonline.org/publications/biographical-memoirs/memoir-pdfs/buchsbaum-solomon-j.pdf

7.42 M. Ya. Azbel' and E. A. Kaner, "Cyclotron Resonance in Metals," *Journal of Physics and Chemistry of Solids* 6 (2–3), 113–135 (August 1958).

7.43 John G. Mavroides (b. 1922), born in Ipswich, Massachusetts, received a BS in electrical engineering from Tufts University in 1944, and MS and PhD degrees from Brown University in 1951 and 1953 respectively. In 1952 he joined Lincoln Laboratory He was a coauthor on at least 16 papers with Ben over the period 1955–1967.

7.44 Mildred "Millie" S. Dresselhaus (1930–2017), who was an Institute Professor Emerita at the time of her death, was hired as a research scientist at Lincoln Laboratory by Ben Lax in June 1960. Her husband, Gene Dresselhaus, was hired at the same time. In those days, Lincoln Laboratory was one of few research laboratories that would hire a married couple. Gene and Millie married in 1958 when they were both at the University of Chicago. Gene's graduate work on cyclotron resonance in Kittel's group at Berkeley was already quite well known to Ben. Millie began her graduate studies in physics at the University of Chicago in 1953. She was influenced by Enrico Fermi, but Fermi died in 1954 and Andrew Lawson was her thesis advisor. Millie would leave Lincoln Laboratory in 1967 for a remarkable academic career at MIT, holding joint appointments in the Physics and the Electrical Engineering departments. She has won numerous awards, including the Buckley Prize in 2008. In recent oral interviews, she referred to Ben as "a great scientist," and gave him credit for diverting her research interests away from superconductivity and toward semimetals, and for giving her the freedom to explore other research paths.

7.45 Morrel H. Cohen, "Energy Bands in the Bismuth Structure. I. A Nonellipsoidal Model for Electrons in Bi," *Physical Review* **121**, 387 (1961).

7.46 The Oliver E. Buckley Prize is awarded annually by the American Physical Society to recognize and encourage outstanding theoretical or experimental contributions to condensed matter physics. Lax received this award in 1960 "for his fundamental contributions in microwave and infrared spectroscopy of semiconductors," which included cyclotron resonance and interband magneto-optical effects. The prize was presented to Lax at the annual dinner of the American Physical Society and the American Association of Physics Teachers in New York City on January 29, 1960 by the president of the American Physical Society, Prof. George Eugene Uhlenbeck (1900–1988) of the University of Michigan. Lax delivered his Oliver E. Buckley Lecture, "Magneto-Spectroscopy in Semiconductors," on March 22, 1960 at the meeting of the American Physical Society in Detroit, Michigan, March 21–24, 1960. Information from Box 5, Folder 28 and Box 6, Folder 24, Benjamin Lax Papers, MIT Archives, and from the Benjamin Lax Files, MIT Museum.

7.46a Peter Wolff, "Matrix Elements and Selection Rules for the Two-Band Model of Bismuth," *Journal of Physics and Chemistry of Solids* **25**, 1057–1068 (1964).

7.47 Peter A. Wolff (1924–2013) was a distinguished theoretical solid state physicist. His career began at Bell Telephone Laboratories (1952–1970). In 1970 he joined the faculty of the MIT Physics Department. In 1981 he succeeded Ben Lax as the second director of the Francis Bitter National Magnet Laboratory, serving until 1987. http://news.mit.edu/2013/peter-wolff-obituary

7.48 This GE paper is most probably the classic 1955 paper by William C. Dash (1926–1963) and Roger Newman, scientists at the General Electric Research Laboratory, Schenectady, New York whose work in the 1950s led to defect-free germanium crystals of high quality: W. C. Dash and R. Newman, "Intrinsic Optical Absorption in Single-Crystal Germanium and Silicon at 77°K and 300°K," *Physical Review* **99**, 1151–1154 (August 15, 1955).

7.49 G. Dresselhaus, A. F. Kip, and C. Kittel, "Cyclotron Resonance of Electrons and Holes in Silicon and Germanium," *Physical Review* **98**, 368–384 (April 15, 1955), received December 16, 1954.

7.50 The Symposium on the Physics of Semiconductors was held at the National Bureau of Standards, Washington, DC, October 24–26, 1956. Ben gave a paper titled "Infrared Investigations with High Pulsed Fields," which immediately followed the paper by E. Burstein of NRL titled "Magnetic Effects on Optical Interband Transitions in InSb." A footnote in the April 1957 *Physical Review* paper by Zwerdling and Lax [P7.15] identifies this conference as "Semiconductor Symposium, October 1956, Washington, DC."

7.51 E. Burstein and G. S. Picus, "Interband Magneto-Optic Effects in Semiconductors," *Physical Review* **105**, 1123–1125 (February 1, 1957), Letter, received December 7, 1956.

7.52 J. M. Luttinger and W. Kohn, "Motion of Electrons and Holes in Perturbed Periodic Fields," *Physical Review* **97**, 869 (1955); J. M. Luttinger, "Quantum Theory of Cyclotron Resonance in Semiconductors: General Theory," *Physical Review* **102**, 1030 (1956).

7.53 The Fourth International Conference on the Physics of Semiconductors was held on August 18–22, 1958, at the University of Rochester, Rochester, New York. W. Crawford Dunlap's article, "The 1958 International Conference on Semiconductors" [*Physics Today* **12** (2), 22 (February 1959)] states that there were 160 papers presented and 490 registered attendees. The proceedings were published as a special issue of *Journal of Physics and Chemistry of Solids* (Vol. 8, pp. 1–552, January 1959). The NRL paper and the Lincoln Laboratory paper appear one after another in the proceedings: E. Burstein, G. S. Picus, R. F. Wallis, and F. Blatt, "Zeeman Type Magneto-optic Studies of Energy Band Structure," pp. 305–311; B. Lax, L. M. Roth, and S. Zwerdling, "Quantum Magneto-Absorption Phenomena in Semiconductors," pp. 311–318.

7.53a The Physical Society held a Conference on Spectroscopy of Solids at the Royal Radar Establishment, Malvern on May 28–29, 1958. The title of Ben's abstract for this conference is "The Theory of Magneto-Absorption Phenomena in Semiconductors." From Box 3, Folder 33 of the Benjamin Lax Papers, MIT Archives.

7.54 The Brussels World Fair, called Expo 58, was held April 17–October 19, 1958. The emphasis of this fair was on science and technology. As part of this fair, the International Conference on Solid State Physics in Electronics and Telecommunications was held in Brussels on June 2–7, 1958. Ben's abstract in the lengthy list of papers has the title "Resonance and Magneto-Absorption in Solids at Millimeters and Infrared Frequencies." From the Benjamin Lax Papers, MIT Archives.

7.55 Professor Robert Allan Smith, CBE, FRS, PRSE (1909–1980), also known as R.A. and as Robin, was a distinguished Scottish mathematician and physicist. He was head of the Physics Department at the Royal Radar Establishment in Malvern, England when Ben visited in the summer of 1958. Smith is remembered for his classic textbooks *Semiconductors* (Cambridge University Press, 1959) and *Wave Mechanics of Crystalline Solids* (Chapman & Hall, 1961), and for his book (with F. E. Jones and R. P. Chasmar), *The Detection and Measurement of Infra-red Radiation* (Oxford University Press, 1957). Between 1962 and 1969, he was Director of the MIT Center for Materials Science and Engineering. https://en.wikipedia.org/wiki/Robert_Allan_Smith

7.56 Pierre Aigrain (1924–2002) was a noted French physicist. He formed a research team in the physics laboratory of the École normale supérieure to do research on semiconductors and other topics in solid state physics. He was head of that team in 1958 when he invited Ben to lecture there.

7.57 Victor Frederick Weiskopf (1908–2002) was an Austrian-born theoretical physicist. He joined the physics faculty at MIT after World War II and was Chair of the Physics Department for some years. Most recently, he was Professor of Physics Emeritus as well as Institute Professor. http://web.mit.edu/physics/people/faculty/weisskopf_victor.html

7.57 Dirk Polder (1919–2001) was a Dutch physicist who, together with Hendrik Casimir, first predicted the existence of what today is known as the Casimir-Polder force, sometimes also referred to as the Casimir effect or Casimir force.

7.58 Hans Joachim Gustav Meyer (1924–2014) was a Jewish German-Dutch theoretical physicist who was working at Philips Laboratories in Eindhoven, Nederland at the time of Ben's visit to that laboratory.

7.59 Cornelis Jacobus Gorter (1907–1980) was a Dutch experimental and theoretical physicist. Among other work, he discovered paramagnetic relaxation and was a pioneer in low temperature physics. In the late 1950s, when Ben visited, Gorter was director of the Kamerlingh Onnes Laboratory in Leiden. One of his doctoral students was Nicolaas Bloembergen, who would receive the Nobel Prize in Physics in 1981.

7.60 Hendrik Brugt Gerhard Casimir *ForMemRS* (1909–2000) was a Dutch theoretical physicist. He is best known for his work (with C. J. Gorter) on the two-fluid model of superconductors in 1934, and for the discovery (with D. Polder) of the Casimir effect in 1948.

7.61 Jay Wright Forrester (1918–2016) was Professor Emeritus in the MIT Sloan School of Management. He founded the field of system dynamics, and was a pioneer of digital computing. He was a key figure in the development of the national air defense system and the MIT Lincoln Laboratory. He led Project Whirlwind, an early MIT digital computing project, which led him in 1949 to invent magnetic core memory, an early form of random access memory. http://news. mit.edu/2016/professor-emeritus-jay-forrester-digital-computing-system-dynamics-pioneer-dies-1119

7.62 Louis Eugène Félix Néel *ForMemRS* (1904–2000) was a French physicist who received the Nobel Prize in Physics in 1970 for his discovery of new phenomena in the magnetic properties of solids, including antiferromagnetism, the opposite of ferromagnetism. The temperature above which this phenomenon ceases is called the Néel temperature. Lax would have visited with Néel at CNRS (Centre National de la Recherche Scientifique), the French National Center for Scientific Research.

7.63 Nicolaas Bloembergen (1920–2017), a Dutch-American physicist, was a professor of physics at Harvard University for most of his academic career. He shared the Nobel Prize in Physics with Arthur Schawlow in 1981, for their work on nonlinear optical phenomena and their application to laser spectroscopy.

7.64 Alan "Al" Louis McWhorter (1930–2018), a native of Crowley, Louisiana, was Professor Emeritus in the MIT Department of Electrical Engineering and Computer Science (EECS) and a scientist and manager at Lincoln Laboratory. He joined Lincoln Laboratory in 1955. He succeeded Ben Lax as Head of the Solid State Division at Lincoln in 1965. McWhorter's ScD thesis for the MIT Department of Electrical Engineering in 1955 is titled "1/f noise and related surface effects in germanium," in which he proposed what has become widely known today as the McWhorter model for 1/f noise as being due to noisy transitions of carriers back and forth between the bulk and traps located at the surface. This model is still the foundation for understanding the origin of 1/f noise in a broad variety of semiconductor devices. http://news.mit.edu/2018/remembering-mit-professor-emeritus-alan-mcwhorter-1121

7.65 Harry Costa Gatos (1921–2015) was born in Greece. He received a PhD in inorganic chemistry from MIT in 1950. His research focus was on the physics and chemistry of electronic materials. His PhD thesis, "Passivity of evaporated iron films in nitric acid," was supervised by H. H. Uhlig.

7.66 Stanley Howard Autler (1922–1991) received his PhD in physics from Columbia University around 1955, after which he joined Lincoln Laboratory. He and Prof. Charles H. Townes are credited with discovering what is called the Autler-Townes effect in 1955, in which a rapidly time-varying electric field induces electronic transitions in atoms.

7.67 Rudolf Kompfner (1909–1977) was an Austrian-born engineer and physicist, who is best known for his invention of the traveling-wave tube in 1943 at the University of Birmingham. He joined Bell Telephone Laboratories in 1951, where he continued development of the traveling-wave tube. https://en.wikipedia.org/wiki/Rudolf_Kompfner

7.68 *High Magnetic Fields, Proceedings of the International Conference on High Magnetic Fields held at the Massachusetts Institute of Technology Cambridge, Massachusetts, November 1–4, 1961*, edited by H. H. Kolm, B. Lax, F. Bitter, and R. G. Mills (MIT Press and John Wiley & Sons, 1962).

7.69 Charles Hard Townes (1915–2015), a native of Greenville, South Carolina, had a remarkable life in science. He shared the 1964 Nobel Prize in Physics with Nikolay Basov and Alexander Prokhorov for the invention of the maser. He received BS and BA degrees from Furman University in Greenville in 1935. He received his MA degree in physics from Duke University in 1937, and his PhD from the California Institute of Technology in 1939. He spent much of his career as professor of physics at Columbia University. Between 1961 and 1967 he was Provost and Professor of Physics at MIT.

7.70 Conference on Quantum Electronics-Resonance Phenomena, held at Shawanga Lodge, High View, New York on September 14–16, 1959. Proceedings were published by Columbia University Press,

New York, 1960. Ben is listed as one of the 11 members of the steering committee, which included Townes of Columbia, Bloembergen of Harvard, Dicke of Princeton, Kittel of UC-Berkeley, and Strandberg of MIT; 163 people attended.

7.71 Robert Hildreth Kingston (b. 1928) received his combined SB and SM degrees in 1948 in the five-year Course 6A Internship Program of the MIT Department of Electrical Engineering, and his PhD in physics from MIT in 1951. His PhD thesis, "A spectroscopic study of the electronic structure of metallic potassium and calcium," was supervised by G. G. Harvey. He spent his career at Lincoln Laboratory and MIT. He wrote two books: *Detection of Optical and Infrared Radiation* (Springer-Verlag, 1978) and *Optical Sources, Detectors, and Systems: Fundamentals and Applications* (Academic Press, 1995). He was elected to the National Academy of Engineering in 1990 for "pioneering quantum electronic device research and its application to modern microwave and optical radar systems."

7.72 Leo Esaki, "New Phenomenon in Narrow Germanium p–n Junctions," *Physical Review* **109**, 603–604 (January 15, 1958), Letter.

7.73 Robert Harmon Rediker (1924–2019) earned his BS degree in electrical engineering in 1947 and his PhD in physics in 1950, both at MIT. His PhD thesis, "Analysis of burst-producing penetrating particles at 10,600 feet," was supervised by Bruno Rossi. Rediker joined Lincoln Laboratory in 1951, where he remained for his entire career. His group developed diffused diodes in GaAs, which enabled Lincoln Laboratory to demonstrate the GaAs laser in 1962. He was also a professor in electrical engineering at MIT.

7.74 R. J. Keyes and T. M. Quist, "Recombination radiation emitted by gallium arsenide diodes," presented at the Solid-State Device Research Conf., Durham, NC, July 1962. R. J. Keyes and T. M. Quist, "Recombination radiation emitted by gallium arsenide," *Proceedings of the IRE* **50**, 1822–1823 (August 1962).

7.75 R. Braunstein, "Radiative Transitions in Semiconductors," *Physical Review* **99**, 1892–1893 (September 15, 1955). This milestone Letter to the Editor by Rubin J. Braunstein (1922–2018) of RCA Laboratories, published just after receiving his PhD in physics from Syracuse University in 1954, is credited with being the first report of an infrared light emitting diode. https://en.wikipedia.org/wiki/Rubin_Braunstein

7.76 R. N. Hall, G. E. Fenner, J. D. Kingsley, T. J. Soltys, and R. O. Carlson, "Coherent Light Emission from GaAs Junctions," *Physical Review Letters* **9**, 366 (November 1, 1962).

7.77 M. I. Nathan, W. P. Dumke, G. Burns, F. H. Dill, Jr., and G. Lasher, "Stimulated Emission of Radiation from GaAs p-n Junctions," *Applied Physics Letters* **1**, 62 (November 1, 1962).

7.78 W. P. Dumke, "Interband Transitions and Maser Action," *Physical Review* **127**, 1559 (September 1, 1962).

Chapter 8   Francis Bitter National Magnet Laboratory, 1958–1981

8.1 George Edward Valley, Jr. (1913–1999) was Professor Emeritus of Physics at MIT. A nuclear physicist by training, he developed critical radar capabilities at MIT during World War II, which led to the creation of Lincoln Laboratory. Prof. Valley was Chief Scientist of the Air Force from 1957–1958, when the proposal for the Magnet Lab was being written and presented, and returned to MIT in 1959. The George E. Valley, Jr. Prize is awarded annually by the American Physical Society to "… one individual in the early stages of his or her career for an outstanding scientific contribution to physics that is deemed to have significant potential for a dramatic impact on the field." http://news.mit.edu/1999/valley-1020

8.2 Henry W. Fitzpatrick (1914–2006) was Assistant Director of Lincoln Laboratory from 1956 until his retirement in 1984. He was responsible for the administration of the laboratory.

8.3 Tokamak is a device that uses a strong magnetic field to confine a hot plasma in the shape of a torus. The tokamak is one of several types of magnetic confinement devices being developed to produce controlled thermonuclear fusion power. As of 2016, it is the leading candidate for a practical fusion reactor. Tokamaks were initially conceptualized in the 1950s by Soviet physicists Igor Tamm and Andrei Sakharov.

8.4 Thomas Howard Stix (1924–2001), Professor Emeritus in astrophysical sciences at Princeton University, was a legendary figure in the field of plasma physics. His research on the heating of plasmas by RF radiation revolutionized the field and opened avenues of approaches toward fusion research. He received a BS in physics from Caltech in 1948 and a PhD in physics from Princeton in 1953. In 1962, he published his classic textbook, *The Theory of Plasma Waves*.

https://pr.princeton.edu/news/01/q2/0417-stix.htm *Physics Today* **55** (3), 94–96 (March 2002).

8.5 The Shubnikov-de Hass effect is oscillations in the low-temperature conductivity *versus* magnetic field of a material, due to Landau levels rising through the Fermi level. It was discovered by Lev Shubnikov and Wander Johannes de Haas.

8.6 Brigadier General Benjamin G. Holzmann (1910–1975) received his BS degree in geology in 1931, his MS degree in geology in 1933, and his PhD in 1934, all from the California Institute of Technology. Between 1958 and 1960, when Ben and company briefed the Air Force, Holtzman was Commander, Air Force Office of Scientific Research, Washington, DC. www.af.mil/About-Us/Biographies/Display/Article/108142/brigadier-general-benjamin-g-holzman/

8.7 Carl F. J. Overhage (1910–1995), a native of London, was a physicist and electrical engineer. He received his BS, MS, and PhD degrees in physics at the California Institute of Technology between 1931 and 1937. He headed a research group at the MIT Radiation Laboratory during World War II. He joined Lincoln Laboratory in 1955. He was Head of Lincoln Laboratory 1957 to 1964, the formative years of the National Magnet Laboratory. In 1961, he became a professor in the MIT Electrical Engineering Department. Carl is editor of *The Age of Electronics* (McGraw-Hill, 1962), a transcription of eight lectures given at MIT during the winter of 1961–1962, celebrating the tenth anniversary of the Lincoln Laboratory. http://news.mit.edu/1995/overhage-0913

8.8 D. Bruce Montgomery (b. 1933) has had a long and distinguished career in the generation and application of high magnetic fields. He participated in the MIT electrical engineering student internship program, earning his combined SB and SM degrees in electrical engineering in 1957. He designed many advanced magnets for the National Magnet Laboratory. He received his ScD degree in electrical engineering from the University of Lausanne.

8.9 The Committee on Perspectives in Materials Research, chaired by Prof. Frederick Seitz of the University of Illinois, was organized in March 1959 and conducted by the Division of Engineering and Industrial Research of the National Academy of Sciences – National Research Council, under contract to the Office of Naval Research, with initial funding at $75,000. Benjamin Lax was one of seven members of the "Techniques and Instrumentation Panel," which was one of at least ten panels comprising this committee. A draft of the final report of Lax's panel, dated August 24, 1959 contains an eight-page section written by Lax and titled "Section VI. High Magnetic Fields," which outlines the new research in materials that would be made possible by having a facility that can provide higher magnetic fields. This material is contained in Folder 7, Box 1 of the Benjamin Lax Papers, MC 234, at the MIT Institute Archives.

8.10   Max Swerdlow (est. 1917–1989) received a BA degree in physics from Brooklyn College in 1938. He then joined the National Bureau of Standards (NBS) as a ceramics engineer. He pioneered the use of the electron microscope at NBS, and eventually headed the NBS laboratory for electron microscopy and electron diffraction. In 1957 he transferred to the Air Force Office of Scientific Research (AFOSR), where he championed the formation of the National Magnet Laboratory and served as the AFOSR program manager for the laboratory until the late 1960s. In his obituary for Swerdlow [*Physics Today* **43** (9), 120–122 (September 1990)], his AFOSR colleague, Dr. Harold Weinstock, said, "*During a 1957 visit to MIT, Swerdlow concluded that the US should establish a national magnet laboratory, built upon the existing facilities in Francis Bitter's lab. Swerdlow promoted this idea to those at MIT (particularly Benjamin Lax) and to his counterparts at other agencies. His efforts were rewarded when, on 1 July 1960, the Francis Bitter National Magnet Laboratory (as it is now known) was formed, with Swerdlow as its very active program manager. After more than a decade in this capacity, he transferred his funding and management responsibilities to the NSF.*"

8.11   This six-member organizing committee for the fledgling National Magnet Laboratory consisted of Francis Bitter (Chairman), Ben Lax (Director), Donald G. Stevenson (Assistant Director), Henry H. Kolm, D. Bruce Montgomery, and James West (Assistant Director for Administration).

8.12   Julius Adams Stratton (1901–1994) earned his SB in 1923 and SM in 1926, both in electrical engineering at MIT. He received his ScD in mathematical physics in 1928 from Eidgenossiche Technische Hochschule, Zurich, Switzerland. He was the 11th president of MIT, 1957–1966, during which period the Magnet Lab was founded. His classic textbook, *Electromagnetic Theory*, was first published in 1941. www.nap.edu/read/12562/chapter/18

8.13 Conrad J. Rauch, "Millimeter Cyclotron Resonance Experiments in Diamond," *Physical Review Letters* 7, 83–84 (August 1, 1961). Rauch was a staff member at Lincoln Laboratory.

8.14 Yuichiro Nishina was a visiting scientist at the Magnet Laboratory who, between 1961 and 1969, collaborated with Ben on seven publications on the Faraday effect. He returned as a professor to Tohoku University.

8.15 W. A. Runciman, from the Physics Department, University of Canterbury, Christchurch, New Zealand, spent a couple of years at the Magnet Lab around 1962–1963, during which time he published several papers on the spectroscopy of rare earth materials.

8.16 Carl Vinton Stager (b. 1935) was a graduate student in the MIT Department of Physics and the Research Laboratory of Electronics. His 1961 PhD thesis, "Hyperfine structure of Hg^{197} and Hg^{199}," was supervised by Francis Bitter. His career was spent in the Department of Physics and Astronomy at McMaster University, Hamilton, Ontario, Canada, where he is Professor Emeritus.

8.17 Hon. Prof. Carl R. Pidgeon, FSRE, is Emeritus Professor of Semiconductor Physics at Heriot-Watt University. He is a Fellow of the Royal Society of Edinburgh and of the Institute of Physics. In some circles, Carl may be better known as father of acclaimed actress Rebecca Pidgeon, who was born in Cambridge, Massachusetts in 1965 while Carl was a Staff Scientist at the Magnet Laboratory.

8.18 Wlodzimierz "Wlodek" Zawadzki (b. 1939) is a theoretical solid state physicist, a writer and a public intellectual in his native Poland. He came to the Magnet Lab in 1965. He is now Professor at the Polish Academy of Sciences in Warsaw. One of the editors (MBR) recalls that one of the first things the young Wlodek did upon arriving in the United States was to send pairs of blue jeans back to his younger brothers in Poland, where such items were a much coveted symbol of America and were unavailable or prohibitively expensive. https://quod.lib.umich.edu/j/jii/4750978.0003.210/--interview-with-wlodzimierz-zawadzki-poland-is-there-life?rgn=main;view=fulltext

8.19 Leonard Sosnowski (1911–1986) was a Polish physicist and a Professor at the University of Warsaw. In the years 1954–1966, he was Director of the Institute of Physics of the Polish Academy of Sciences, where his focus was on the physics of semiconductors.

8.20 George Buford Wright (1926–2017) received his PhD in physics at MIT in 1960. His thesis title was "Magneto-optical properties of mercury selenide." His thesis supervisor was Prof. Arthur Robert von Hippel. Wright joined Lincoln Laboratory around 1960. He later joined the physics faculty at Stevens Institute of Technology, and then was at the Office of Naval Research until his retirement.

8.21 C. R. Pidgeon and R. N. Brown, "Interband Magneto-Absorption and Faraday Rotation in InSb," *Physical Review* **146**, 575–583 (June 10, 1966).

8.22 Luis Walter Alverez (1911–1988) was a legendary physicist with broad and deep accomplishments. He worked at the MIT Radiation Laboratory during World War II, where he was the primary inventor of the Ground Control Approach radar system for the blind landing of aircraft. He then worked on the atomic bomb at Los Alamos. He did experimental physics on cyclotrons, particle accelerators and bubble chambers at UC–Berkeley with Ernest Lawrence. He used cosmic rays to "X ray" an Egyptian pyramid, and developed a new theory about the extinction of the dinosaurs. He won the 1968 Nobel Prize in Physics for his work on elementary particles. www.nasonline.org/publications/biographical-memoirs/memoir-pdfs/alvarez-luis-w.pdf

8.23 Major General James McCormack, Jr. (1910–1975), a native of Chatham, Louisiana, graduated West Point in 1932. He later earned a master's degree in civil engineering at MIT in 1937. He served in the United States Atomic Energy Commission (AEC) to oversee research and development of nuclear power and nuclear weapons. In 1950, he transferred to the Air Force, from which he retired in 1955 to head the Institute for Defense Analyses. In 1958, around the time of early planning of the Magnet Laboratory, he became Vice President for industrial and governmental relations at MIT.

8.24 Richard Freeman Post (1918–2015) earned his PhD in physics from Stanford University in 1951. He spent his entire remarkable scientific career at Lawrence Livermore National Laboratory advancing fusion science and technology. *Physics Today* **68** (7), 56 (July 2015).

8.25 Richard "Dick" Neil Brown (1931–2018) joined Lincoln Laboratory as a research physicist in 1956. He earned his SM degree in physics at MIT in 1958. His thesis title is "Measurements of effective masses of charge carriers in semiconductors by Faraday rotation." His thesis supervisor is listed as Prof. George Fred Koster, but Lax, who was not yet on the physics faculty, was the unofficial supervisor. Dick transferred to the Magnet Lab in the early 1960s. He left the Magnet Lab in 1967 for the US Air Force Geophysics Laboratory at Hanscom Air Force Base, Bedford, Massachusetts. He retired in 1993.

8.26 Evan O'Neill Kane (1924–2006) was a theoretical physicist who in 1957, while working at the General Electric Research Laboratory in Schenectady, New York published what has become a landmark paper on the mathematical description of the energy band structure in semiconductors, referred to as the **k·p** approximation. This paper has been cited over 3,300 times. E. O. Kane, "Band Structure of Indium Antimonide," *Journal of Physics and Chemistry of Solids* **1** (4), 249–261 (1957).

8.27 Prof. Jacek Furdyna, Department of Physics, University of Notre Dame, was a Staff Member at the National Magnet Laboratory from 1962 to 1966.

8.28 There were eight rooms or "cells" on the main floor of the Magnet Lab. Each cell had two Bitter magnets, but only one in each could be operated at one time. There were two larger rooms where specialized magnets were stationed and operated. There were three shifts each day, morning, afternoon, and evening, each lasting about four hours. At the beginning of a shift, the cooling water pumps would be connected to the selected magnets and turned on. There would be much noise and vibration of the sturdy rubber hoses carrying the cooling water into and out of the magnet, even though the hoses were lashed down. The copper bus bars would be switched into place to provide current to the selected magnets. Then the user could control the current level through a control box connected to the control room and proceed with his experiments. There was a red panic button that would quickly reduce the magnetic field to zero in an emergency. Before the current was turned on, the area around the magnet was carefully inspected for any loose tools or other magnetic material. A loose screwdriver could turn into a missile in the high magnetic fields. In those days we used the large Tektronix oscilloscopes with cathode ray tubes, and it was always fascinating to see the trace on the display twist when the magnetic field was turned on.

8.29 N. Miura, G. Kido, and S. Chikazumi, "Infrared Cyclotron Resonance in InSb, GaAs and Ge in Very High Magnetic Fields," *Solid State Communications* **18** (7), 885–888 (1976).

8.30 Norman Allen Blum (b. 1932) received his AB at Harvard University in 1954 and his PhD at Brandeis University in 1964. He was a Lieutenant in the US Navy, 1954–1957. https://prabook.com/web/norman_allen.blum/3336146

8.31 Richard B. Frankel is an Emeritus Professor of Physics at the California State Polytechnic University, San Luis Obispo. He earned his PhD from the University of California at Berkeley in 1965. He was at the Magnet Lab from 1965 to 1988. https://en.wikipedia.org/wiki/Richard_B._Frankel

8.32 R. H. Rediker and A. R. Calawa, "Magnetotunneling in Lead Telluride," *Journal of Applied Physics* **32** (10), 2189–2194 (November 1, 1961).

8.33 J. F. Butler, "Magnetoemission Experiments in $Pb_{1-x}Sn_xTe$," *Solid State Communications* **7** (13), 909–912 (1969). Butler was a staff member at Lincoln Laboratory.

8.34 Robert Gerson Shulman (b. 1924) is a biophysicist who worked at Bell Labs from 1953 to 1979. He was a pioneer in the application of nuclear magnetic resonance (NMR) to study biochemical processes.

8.35 Leo J. Neuringer (1928–1993), an alumnus of Rensselaer Polytechnic Institute, received his PhD from the University of Pennsylvania in 1957, after which he joined Raytheon. In 1963, he joined the MIT Francis Bitter National Magnet Laboratory, where he spent the rest of his career. In the mid-1970s, he became the champion for NMR research and related activities, including NMR imaging and biomedical applications. He founded and directed the Magnet Laboratory's High Field Nuclear Magnetic Resonance Resource in 1974 and its Magnetic Resonance Imaging Facility in 1982. *Physics Today* **47** (1), 60 (January 1994). http://news.mit.edu/1993/neuringer-0512

8.36 Robert Guy Griffin (b. 1942) a native of Little Rock, Arkansas, is a professor in the MIT Department of Chemistry. Since 1992, he has been Director of the Francis Bitter Magnet Laboratory. He received his PhD in 1969 from Washington University, Saint Louis, Missouri. He then joined Prof. John Waugh in the MIT Chemistry Department as a postdoctoral student, and joined the Magnet Lab in 1972. His research areas are nuclear magnetic resonance and high-field dynamic nuclear polarization in biological solids.

8.37 Charles Elroy Chase, Jr. (est. 1929–2007) received his BS in physics at MIT in 1950, and his PhD from the University of Cambridge in 1954. In addition to a research physicist, Charlie was also a talented pianist and composer.

8.38 Robert Hilton Meservey (1921–2013) joined Lincoln laboratory in 1961 after receiving his PhD in low-temperature physics from Yale University. He moved to the Magnet Lab in 1963, where he had a long and productive research career, leading a group in low temperature physics. In 1970 he and Paul M. Tedrow discovered electron spin-polarized tunneling. In 2009, Meservey and three colleagues, Jagadeesh Moodera and Paul M. Tedrow, both of the Magnet Lab, and Terunobu Miyazaki of Tohoku University, Japan, received the Oliver E. Buckley Condensed Matter Physics Prize from the American Physical Society for their "pioneering work in the field of spin-dependent tunneling and for the application of these phenomena to the field of magnetoelectronics." http://news.mit.edu/2013/obit-meservey-physics https://physicstoday. scitation.org/do/10.1063/PT.5.6007/full/

8.39 Emanuel "Manny" Maxwell (1913–2000) received his PhD in physics from MIT in 1948. His PhD thesis, "The surface impedance of tin at 24,000 Mc/sec in the superconducting and normal states," was supervised by John C. Slater. After working on radar at the MIT Radiation Laboratory during World War II, he joined Lincoln Laboratory. In 1963 he joined the Magnet Lab, were he worked until his retirement in 1983. http://news.mit.edu/2000/maxwell-1018

8.40 H. A. Gebbie was a visiting scientist at the Magnet Lab from the National Physical Laboratory, Teddington, England. In 1965 Ken Button convinced Gebbie to bring to the Magnet Lab a copy of the newly discovered cyanide laser operating at a wavelength of 0.337 mm. Using that laser, Button, Gebbie and Lax studied cyclotron resonance in p-type germanium at magnetic fields up to 180 kilogauss and temperatures down to 40 K.

8.41 David M. Larsen received his BS in physics from MIT in 1957, and his PhD in physics from MIT in 1962. His PhD thesis, "Theory of mixtures of dilute hard sphere gases at low temperature," was supervised by Paul Federbush. Dave spent his career at Lincoln Laboratory and the Magnet Laboratory.

8.41a Quirin H. F. Vrehen (b. 1932) was born in Hertogenbosch, The Netherlands. He received his PhD in physics from the State University of Utrecht in 1963. His thesis dealt with electron spin resonance and optical properties of solids. Between 1963 and 1966 he was a staff member at the MIT National Magnet Laboratory, where he did magneto-optical experiments on semiconductors in crossed and parallel electric and magnetic fields. He then joined the Philips Research Laboratories, Eindhoven, and later taught at the University of Leiden. He has been a member of the Royal Netherlands Academy of Arts and Sciences since 1988.

8.42 Daniel Ross Cohn (b. 1943) was an MIT physics graduate student of Ben's. After receiving his PhD in 1971 [T11], Dan joined the Magnet Laboratory, where he collaborated closely with Ben on wide-ranging research in plasmas and gaseous lasers. Dan became a close friend, colleague, coauthor, and skiing partner of Ben's. Later, Dan joined the MIT Plasma Science Fusion Center, where he was a Senior Research Scientist and Head of the Plasma Technology Division. After retiring from the Fusion Center and from the MIT Nuclear Engineering Department, he joined the MIT Energy Initiative as a Research Scientist.

8.43 David Cohen joined the Magnet Lab in 1969, with the objective of building a magnetically shielded room that would allow the tiny currents from the human heart and brain to be detected and imaged. His outstanding research efforts over the succeeding decades proved to be enormously successful and transformative, with major clinical applications. Cohen is often referred to as the father of biomagnetism. He pioneered the new fields of magnetoencephalography (MEG) and magnetocardiography (MCG). Today, there are over 200 MEG machines in use around the world. http://davidcohen.mit.edu/

8.44 Arthur H. Guenther (1931–2007), a native of Hoboken, New Jersey, had a long career with the US Air Force, spending 31 years at Kirtland Air Force Base, Albuquerque, New Mexico. For much of that tenure he was Chief Scientist of the Air Force Weapons Laboratory, which is now part of the Air Force Research Laboratory. He received a BS in chemistry from Rutgers University, and received his PhD in chemistry and physics from Pennsylvania State University in 1957.

8.45 Lowell Lincoln Wood, Jr. (b. 1941) is an American astrophysicist who has worked with the Strategic Defense Initiative and geoengineering. His PhD in 1965 in geophysics is from the University of California at Los Angeles. Wood has over 1,780 patents, surpassing Thomas Edison.

8.46 Richard Douglas Thornton, retired since around 1997, was a professor in the MIT Electrical Engineering and Computer Science Department for 40 years. He earned his ScD from MIT in 1957. His thesis, "Some limitations of linear amplifiers," was supervised by Prof. Henry J. Zimmermann. Richard helped found MagneMotion Inc. He has worked on magnetic levitation and linear motor development since 1968. He participated in the National Maglev Initiative and the Urban Maglev Project. His early work with Henry Kolm on the magneplane at the Magnet Lab formed part of the foundation of Richard's career.

8.47 The Fractional Quantum Hall Effect was discovered at the Francis Bitter National Magnet Laboratory by Horst L. Störmer and Daniel Chee Tsui on October 7, 1981 using a 23 tesla (230 kilogauss) Bitter magnet. For this discovery, they and Robert B. Laughlin were awarded the 1998 Nobel Prize in Physics.

8.48 Hernan Camilo Praddaude (1932–2018), a native of Rosario, Argentina, earned his PhD in physics at MIT in 1964. Francis Bitter supervised his thesis, "Density matrix formalism: Application to the theory of optical pumping." He joined the Magnet Lab in 1963.

8.49 In August 1990, in a competitive procurement, NSF decided to award a $60M five-year contract to Florida State University, Tallahassee, Florida, to construct from scratch a new national high magnetic field facility, rather than to the incumbent, MIT, to improve the existing Francis Bitter National Magnet Laboratory. "Coming Attraction: NSF Rejects MIT, Picks Florida State for Magnet Lab," *Physics Today* 44 (1), 53–56 (January 1991).

Chapter 9 Professor of Physics at MIT, 1965–1981

9.1 Walter Gordy (1909–1985), a native of rural Newton County, Mississippi, was the James B. Duke Professor of Physics at Duke University. He is called one of the founding fathers of microwave spectroscopy. He was at the MIT Radiation Laboratory in 1942–1946 before going to Duke. He founded and directed the Duke Microwave Laboratory. He is a member of the National Academy of Sciences. It is clear why Prof. Gordy would have greatly desired that Ben join his group. www.nap.edu/read/12042/chapter/7 https://phy.duke.edu/about/history/historical-faculty/walter-gordy

9.2 George Bernard Benedek (b. 1928) was a Professor in the MIT Physics Department when Ben's appointment was being finalized. Benedek taught a graduate course in solid state physics for several years thereafter. He received his PhD in physics from Harvard University in 1953. He is Professor Emeritus of Physics and Biological Physics in the MIT Department of Physics, and Professor of Physics and Health Sciences and Technology in the Harvard–MIT Division of Health Sciences and Technology.

9.3 A. Mooradian and G. B. Wright, "Observation of the Interaction of Plasmons with Longitudinal Optical Phonons in GaAs," *Physical Review Letters* **16**, 999–1001 (May 30, 1966).

9.4 Timothy Richard Hart (1940-1994) received his PhD in physics from MIT in 1970. His thesis, "Phonons and magnons in solids," was supervised by Benjamin Lax. After receiving his PhD, Tim joined the physics faculty at Stevens Institute of Technology, Hoboken, New Jersey, where he remained until his untimely death at the age of 54. He held various positions at Stevens. He was appointed Head of the Department of Physics & Engineering Physics in 1989. He also served as Dean of the Graduate School in 1991-1993. He was named Dean of Research in 1993. Lax remembers Tim as a gifted experimentalist and an enthusiastic mentor for younger graduate students at the Magnet Laboratory. https://physicstoday.scitation.org/do/10.1063/PT.6.4o.20190128a/full/

9.5 R.A. Cowley, "Raman Scattering from Crystals of the Diamond Structure," *Journal de Physique France* **26** (11), 659–667 (November 1965).

9.6 Margaret Horton Weiler (b. 1940) was hired by Ben in the fall of 1965 to work with Ken Button at the Magnet Lab to set up experiments with two lasers that she built for him (CO_2 and HCN). Ben soon changed her status from half-time NML staff member to full-time MIT graduate student. She earned a PhD in physics in 1977, and then joined the MIT Physics Department faculty. She recalls that Ben "… was an important part of my career, including encouraging me to stay working after my two children were born – half-time and mostly from home – on theoretical studies of narrow-gap semiconductors such as HgCdTe, InSb, and Te. I was able to do most of the work at home using a teletype connected to the MIT computers using their new remote access capability. Ben was one of the early supporters of this and other new capabilities." She would go on to have a distinguished and productive career in industrial physics, developing advanced HgCdTe infrared detectors for military and space applications.

9.7 Second International Congress on Waves and Instabilities in Plasmas, Innsbruck, Austria, March 17–21, 1975.

9.8 Winter Olympics XII were held in Innsbruck, Austria, February 4–15, 1976.

9.9 Dan Cohn, a 1971 MIT PhD graduate student and a longtime friend of Ben's, as well as a frequent skiing companion, recalls their Austrian skiing adventure: "It was a complete white out. You could not see beyond 50 feet or so. And we were on changing terrain with some sharp drops. Although Ben had just started skiing, he was very calm and determined, and we just pressed on until we got to the bottom."

Chapter 10   Emeritus Years and Consulting, 1981–2006

10.1 William Eugene Keicher (b. 1947) earned his BS (1969), MS (1970), and PhD (1974) in electrical engineering at Carnegie-Mellon University. He joined Lincoln Laboratory in 1975. His career has centered on coherent laser radar technology.

10.2 Joshua Zak (b. 1929) is a theoretical physicist and a professor in the Department of Physics, Technion – Israel Institute of Technology, Haifa, Israel. In 1961, Lax invited Zak to the Magnet Lab for the academic year 1963–1964, where he discovered the Magnetic Translation Group. He then returned to the Technion as a faculty member, where he has remained ever since. In 1967, Joshua was invited again to MIT as visiting professor, where he discovered the "Zak Transform." He has received numerous awards for these major contributions to condensed matter physics.

10.3 H. Angus MacLeod, *Thin Film Optical Filters*, fifth edition (CRC Press, 2017). This classic textbook and reference work was first published in 1969.

Biographical References for Prof. Benjamin Lax

1. Benjamin Lax, "Biographical Note," pp. 123–124 in "The effect of magnetic field on the breakdown of gases at high frequencies," PhD Thesis, Physics, MIT 1949.
2. *Five Years at the Radiation Laboratory*, internal publication of the Massachusetts Institute of Technology, 1946. MIT Archives and Special Collections.
3. Noah Gordon, "Dr. Lax Uncovers World of Crystals: Hungary-Born Scientist Had Passion for Math," *Boston Sunday Herald*, February 21, 1960.
4. Interview of Benjamin Lax by Joan Bromberg on May 15, 1986, Niels Bohr Library & Archives, American Institute of Physics, College Park, Maryland. This interview focusses on research on the solid-state maser and the semiconductor maser and laser at the MIT Lincoln Laboratory in 1963. www.aip.org/history-programs/niels-bohr-library/oral-histories/4735
5. Benjamin Lax, an oral history conducted on June 13, 1991 by Frederik Nebeker, IEEE History Center, Hoboken, New Jersey. http://ethw.org/Oral-History:Benjamin_Lax. Nebeker was a senior research historian at the IEEE History Center at Rutgers University. His interview covers Ben's experiences in the Army and at the MIT Radiation Laboratory during the years 1942 to 1949.
6. Stuart W. Leslie, *The Cold War and American Science: The Military-Industrial-Academic Complex at MIT and Stanford* (Columbia University Press, 1993). Pages 195–201 contain a brief history of the MIT National Magnet Laboratory.

7. *MIT Lincoln Laboratory: Technology in the National Interest*, Eva C. Freeman, ed. (MIT Lincoln Laboratory, 1995). Pages 194–199 contain a history of the early days of solid state physics and device research, 1954–1963, written by Lax and covering ferrites, cyclotron resonance, and magneto-optical spectroscopy.

8. "Benjamin Lax memoir 1998–2000," interviews with Benjamin Lax conducted by Donald T. Stevenson, 1998–2000, unpublished. Original audio tapes and raw transcripts were donated in 2005 to the Niels Bohr Library & Archives, American Institute of Physics, College Park, Maryland.

9. "Benjamin Lax (1915–2015), A Biographical Memoir" by Roshan L. Aggarwal and Marion B. Reine, published online by the National Academy of Sciences, March 2016. www.nasonline.org/publications/biographical-memoirs/memoir-pdfs/lax-benjamin.pdf

10. Daniel Cohn, "Professor Emeritus Benjamin Lax dies at 99: Pioneer in semiconductors created MIT's magnet laboratory and served as associate director of the MIT Lincoln Laboratory," *MIT News*, April 27, 2015. http://news.mit.edu/2015/professor-emeritus-benjamin-lax-dies-0427

11. Richard Temkin and Daniel Cohn, "Benjamin Lax," *Physics Today* **69** (3), 66 (March 2016), Obituary.

12. Benjamin Lax online biography, American Institute of Physics, Physics History Network. https://history.aip.org/phn/11604023.html

13. Joseph Daniel Martin, "Solid Foundations: Structuring American Solid State Physics, 1939–1993," PhD Thesis, University of Minnesota, Department of Philosophy, May 2013. In his thesis and in his book listed below, Martin treats the Francis Bitter National Magnet Laboratory as a case study of a major research laboratory devoted largely to solid state physics.

14. Joseph D. Martin, *Solid State Insurrection: How the Science of Substances Made American Physics Matter* (University of Pittsburgh, Press, 2018). Chapter 5, "Big Solid State Physics at the National Magnet Laboratory."

15. Benjamin Lax Papers, MC 234, Institute Archives and Special Collections, Massachusetts Institute of Technology, Cambridge, Massachusetts.

16. Benjamin Lax Files, MIT Museum Archives.

Acknowledgments

W E ARE INDEBTED TO many people and institutions for their help with this book. Dr. Margaret H. Weiler, Prof. Dennis L. Polla, and Dr. Peter Capper reviewed the original proposal for this book. Dr. Daniel R. Cohn shared recollections of his skiing adventures with Ben. Prof. Jay L. Hirshfield of Yale University provided recollections of the electron cyclotron laser experiments done at the Magnet Laboratory. Myles Crowley at the MIT Archives and Special Collections and Rachael Robinson at the MIT Museum provided valuable assistance. Dr. Ward D. Halverson provided encouragement and located key references. The staff at the Harvard University Archives were most helpful. Daniel R. Lax provided several photographs of his father. Edmund K. Summersby retrieved useful data from the Harvard Alumni Office. Barbara Walsh of the *Microwave Journal* provided references to Ben's editorials and articles in that journal. Angela Locknar of the MIT Lincoln Laboratory Research Library was most helpful in tracking down several elusive references. Dr. Thomas Miller of Boston College Institute for Scientific Research provided the abstracts of Ben's presentations at the 1949, 1950, and 1951 Gaseous Electronics Conferences. Kathleen Reine did a critical read of the final manuscript.

Several people provided permissions to use the photographs that appear in this book: Katie Zimmerman of the MIT Libraries, Prof. Nan Jackson and Mark W. Jackson, Sherry E. Aaron, and Fabian Bachrach.

We thank our editors at CRC Press/Taylor & Francis, Rebecca Davies and Dr. Kirsten Barr, for expertly guiding us through the publication process. We thank the production team, Linda Leggio at CRC Press/Taylor & Francis and Sarah Green at Newgen Publishing UK, for their diligent efforts. And we thank Claire Bell for her careful and faithful copyediting.

About the Interviewer and Editors

Dr. Donald T. Stevenson (Interviewer) was a close colleague and friend of Ben's throughout their long careers at MIT. Don received his PhD in physics at MIT in 1950. He joined MIT Lincoln Laboratory in 1951, the same year as Ben. He assisted Ben in the planning, construction and operation of the MIT National Magnet Laboratory. He retired from MIT in 1988 as Assistant Director of the Francis Bitter National Magnet Laboratory.

Dr. Marion B. Reine (Editor) is a former MIT PhD physics graduate student of Ben's. After a 40-year career as an industrial physicist, advancing the physics and technology of semiconductor quantum infrared detectors, he retired in 2010 and is now a consultant on infrared detectors to industry, government, and academia. He has over 160 publications and reports, and is a Fellow of the American Physical Society.

Dr. Roshan L. Aggarwal (Editor) joined MIT National Magnet Laboratory in 1965, where he worked closely with Ben on the optical properties of semiconductors for nearly 30 years. He was Associate Director of the Francis Bitter National Magnet Laboratory during 1977–1984. He retired from Lincoln Laboratory in 2016 after 51 years of research and teaching at MIT. He has since published two books: *Introduction to Optical Components*, coauthored with K. Alavi (CRC Press, 2018), and *Physical Properties of Diamond and Sapphire*, coauthored with his PhD Thesis Advisor at Purdue, Prof. A. K. Ramdas (CRC Press, 2019). A third book, *Simple Experiments in Optics*, coauthored with K. Alavi, was published in 2019 by Cambridge Scholars Publishing.

Index

Note: Page numbers in *italics* refer to figures.